The New Mobility Handbook

2024 Editior

The New Mobility Handbook

2024 Edition

MICHELE KYROUZ

Warrendale, Pennsylvania, USA

400 Commonwealth Drive
Warrendale, PA 15096-0001 USA
E-mail: CustomerService@sae.org
Phone: 877-606-7323 (inside USA and Canada)
724-776-4970 (outside USA)
FAX: 724-776-0790

Copyright © 2024 by Michele Kyrouz. All rights reserved.

No part of this publication may be reproduced, stored in a retrieval system, or transmitted, in any form or by any means, electronic, mechanical, photocopying, recording, or otherwise, without the prior written permission of SAE International. For permission and licensing requests, contact SAE Permissions, 400 Commonwealth Drive, Warrendale, PA 15096-0001 USA; e-mail: copyright@sae.org; phone: 724-772-4028.

Library of Congress Catalog Number 2023950773
http://dx.doi.org/10.4271/9781468607086

Information contained in this work has been obtained by SAE International from sources believed to be reliable. However, neither SAE International nor its authors guarantee the accuracy or completeness of any information published herein and neither SAE International nor its authors shall be responsible for any errors, omissions, or damages arising out of use of this information. This work is published with the understanding that SAE International and its authors are supplying information but are not attempting to render engineering or other professional services. If such services are required, the assistance of an appropriate professional should be sought.

ISBN-Print   978-1-4686-0707-9
ISBN-PDF   978-1-4686-0708-6
ISBN-ePub   978-1-4686-0709-3

To purchase bulk quantities, please contact: SAE Customer Service

E-mail: CustomerService@sae.org
Phone: 877-606-7323 (inside USA and Canada)
724-776-4970 (outside USA)
Fax: 724-776-0790

Visit the SAE International Bookstore at books.sae.org

Publisher
Sherry Dickinson Nigam

Product Manager
Amanda Zeidan

Production and Manufacturing Associate
Michelle Silberman

For Will and Kate
You make everything possible.
Thank you for your love, encouragement, and endless good cheer.
~

Contents

Introduction   x

PART I
Change How We Use Cars   1

CHAPTER 1
Unbundle the Car   3

Why We Like Cars   3
The Right Tool for the Job   4
One Ride at a Time   5
Regulate All Cars   7

CHAPTER 2
Price Road Use Fairly   11

Pay by the Mile   12
Traffic Is Caused by Demand   13
Surge Pricing Will Reduce Traffic   16

CHAPTER 3
Reallocate Space on Our Streets — 23

Parking Is for Garages — 24
Protected Lanes for Micromobility — 25
Let Buses Run Free — 27
Policy Change Is Possible — 27

PART II
Make Micromobility Work — 31

CHAPTER 4
Invest in Infrastructure — 33

Micromobility Is Good for Cities — 33
Micromobility Needs Infrastructure — 37
A Place to Park — 38
A Place to Ride — 41
We'll Always Have Pari- — 45

CHAPTER 5
Reset the Scooter Rules — 49

Scooter Economics — 49
Number of Operators — 50
Fleet Size and Parity — 53
Operating Areas — 55
Permit Length — 57
Selection Criteria — 58
Regulatory Burden — 59
Permit Fees — 60
Labor Costs — 62
Equity Zones — 63
Parking Compliance and Fines — 64

Rider Restrictions	66
Sidewalk Riding	68

PART III

Make Public Transit Great — 71

CHAPTER 6

Fix Buses First — 73

The Public Transit Struggle	73
Don't Make Other Modes Worse	74
Fast, Frequent, and Free-ish	75
Bus Lanes	77
Signal Priority	80
Boarding/Payment	81

CHAPTER 7

Plan for the Future — 85

Mobility on Demand	85
The Future Will Be Automated	89
First-Mile/Last-Mile Solutions	91
A Fairer Fare?	93
Index	97
About the Author	101

Introduction

New mobility technologies and urbanism have been at odds since Uber ruffled feathers in cities over a decade ago. In the years since the pandemic, which decimated public transit systems, the conflict has gotten worse. Many city transportation planners believe that new mobility technologies from autonomous vehicles to micromobility only serve to detract from public transit and should be discouraged.

New mobility options are incredibly popular and can encourage multimodal travel in ways that public transit has not. Together, new mobility and public transit can provide viable alternatives to personal car use. What has been missing to date is for cities to use classic urban planning principles such as road pricing and reallocation of road space to mitigate the negative externalities that new mobility options might otherwise cause, rather than trying to ban or hinder them. Ride services and micromobility are making these policies more attractive. Efforts to reshape our streets, reduce parking for cars, implement road pricing, and install dedicated bus and bike lanes become more acceptable as more people want to use new mobility options. New modes bring new riders and new advocates for policies that encourage multimodal travel.

With ride services available on demand, the introduction of autonomous ride services in cities such as San Francisco and Austin, and more electric bikes and scooters approved, the promise of new technologies has shifted the political calculus in support of multimodal transportation. Public transit has not succeeded in disrupting car dominance by itself, and new strategies are needed. Even after 50 years of "transit first" policies, efforts to reduce car travel have failed, largely due to a lack of attractive alternatives to the private car. In the United States (US), public transit accounts for fewer than 2% of passenger trips.[1] Cars are still the dominant mode of transportation for every trip from one to 100 miles. This has been true for decades, long before any new mobility technologies emerged. The challenge of disrupting cars is one that new mobility companies and urbanists can only solve together—by allowing new modes to operate and changing how we charge for and allocate space to private cars.

This book advocates for cities to work with new mobility providers to bring together the smarter use of cars, micromobility, and transit as part of a combined ecosystem rather than pitting one against the other. Part I argues that we should change how we use cars in cities with fair pricing and allocation of street space for all modes. Part II makes the case for micromobility as a vital part of city transportation. Part III explains why making public transit a great option, not a last resort, is important even for those who choose not to ride the bus. Cities need all of these options to work together to create a new mobility system for the next century.

---

[1] U.S. Department of Transportation, Bureau of Transportation Statistics, "Passenger Travel Facts and Statistics 2016," p. 11 (most recent data from 2009 show that transit is 1.9% of trips and 1.5% of miles).

Part I
Change How We Use Cars

Erosion of cities by automobiles entails so familiar a series of events that these hardly need describing. The erosion proceeds as a kind of nibbling, small nibbles at first, but eventually hefty bites...More and more land goes into parking, to accommodate the ever increasing numbers of vehicles while they are idle.[1]

—*Jane Jacobs*

---

[1] Jane Jacobs, *The Death and Life of Great American Cities* (Vintage Press, 1961), p. 349.

© 2024 by Michele Kyrouz

Marc Sitkin/Shutterstock

1

Unbundle the Car

Why We Like Cars

For 60 years, the most convenient mode of transportation has been driving your own car. Even as problems with cars in cities became evident, we had no ready solution or reasonable substitute for the car. It turns out we cannot begin the "attrition" of automobiles that Jane Jacobs suggested without offering attractive alternatives that people want to use.[1] We often criticize car culture as wasteful, polluting, and inefficient. Yet, most people commute to work in their own cars every day. With the average car in the US weighing about 4000 to 5500 lb and transporting just one person, this is an enormous waste of space and energy. Why do we spend so much money and energy to drive huge vehicles taking up space and sitting in traffic?

It turns out that driving your own car is incredibly convenient and comfortable. It is this fundamental truth that city planners, public transit proponents, urbanists, and bicycle advocacy groups have failed to address. We like our cars because they provide three benefits:

- Point-to-point travel
- With no fixed schedule
- In a comfortable, private environment

---

[1] Jane Jacobs, *The Death and Life of Great American Cities* (Vintage Books, 1961), p. 349.

What other mode of travel did we create in the last 100 years that can compare? For decades, there has been no reasonable alternative to owning and driving your own car.

Unlike public trains and buses, a car takes you directly from your home to wherever you need to go and whenever you need to get there. If you need to step out of work for a doctor's appointment or a concert at your child's school, your car stands ready to take you. Your car has the car seats you need already installed with toys and diaper bags at the ready. Your car offers privacy to take work calls during the commute. For women and the elderly in particular, your car offers relative safety, where you do not have to be constantly vigilant against possible threats or petty crime.

Given the comfort and convenience of driving your own car, getting Americans to try different modes of travel has been completely unsuccessful. This is particularly true because car rides required car ownership for most Americans. Owning your own car has economic deterrents to trying and using other modes of transportation. We do not yet know whether ride services such as Uber, Lyft, Waymo, and Cruise can change that paradigm—whether *not* owning a car will lead to more multimodal trips. We do have over 60 years of undisputed evidence that when most people own cars they do not use other modes.

The Right Tool for the Job

Not all trips require a car. Yet once people buy cars, they use them for every trip they take—from 1 to 100 miles. After all, the car is already in the driveway. Sometimes a car is the right tool for the job, but often in cities it is entirely unnecessary. Therefore, it makes sense to encourage multimodal travel by starting with the premise that if we make car rides available one at a time and not only through personal car ownership, people might make different choices.

If you do not buy a car, you can determine for each trip you take what the "right tool for the job" might be. Analyst Horace Dediu has called this the "unbundling of the car." He describes car ownership as a "bundle of trips" that you pay for upfront.[2] When you buy a car, you are buying all the trips you will take in the car in the future and you are incentivized to use it for all trips, whenever possible, since you have already paid for it and continue to incur costs to take care of it.[3] This sunk cost drives many of our transportation decisions. Each new trip has only the marginal cost of gasoline and some wear and tear, rather than the full cost of the trip built in, as it would if you purchased each ride separately with Uber or Lyft, or a carsharing service.

By unbundling the car from an asset you own into a series of rides you buy one at a time, we can reduce the number of trips in cars and encourage the use of other modes where appropriate. Perhaps for a one-mile trip in a downtown area, it would be faster and more fun to ride a shared electric scooter that can be dropped off at your destination. Untethered from your car as a default option, you are open to considering other modes as you move around the city.

The unbundling of the car encourages multimodal travel, but most of us will still need to take many trips in cars. Car travel is, and will continue to be, a key mode in multimodal travel. This is why ride services are a crucial piece of the puzzle for encouraging multimodal travel. With a car ride available at the push of a button, we can have the benefit of *riding* in a car without the hassle of *owning, driving, and parking* one.

Some urbanists call for a "war on cars," but this approach overstates both the problem and the solution. We do not need to ban cars, just change how and when we use them. This is an important distinction that calls for a more nuanced set of solutions.

One Ride at a Time

Since the 1950s, car ownership has been key to mobility of another type—access to jobs, housing, and the American dream. This has stemmed from the relative unavailability of car rides without car

---

[2] Horace Dediu, "The Car Will Be Unbundled," *Micromobility Industries - Our Vision*, accessed at https://micromobility.io/our-vision.
[3] Horace Dediu, "Part 2: Disruption," *Micromobility Industries*, Jan. 22, 2019.

ownership, even in big cities. Taxis have historically been very expensive and hard to order and rely on in most US cities other than New York. For those in outer neighborhoods, especially neighborhoods of color or with lower socioeconomics, it was virtually impossible to get a taxi ride when needed. For elderly people or those with disabilities preventing driving, the inability to get a reliable ride created similar disadvantages, including lack of opportunity and isolation. Before working from home was a norm, the lack of ability to drive or get a ride severely limited job options.

The most important new mobility technology in 100 years was not a new form factor: It was an app on your phone giving you the ability to get a car ride on demand in minutes, without owning, driving, and parking a car. The rise of ride services available on demand changed the fundamental paradigm that *car ownership* was necessary in order to reliably get a car ride when needed. Ride services have improved mobility for many groups, including young people who want to go out for drinks across town, elderly and disabled people who were reliant on others for rides, people in transit-sparse neighborhoods who need a ride to work, and those who cannot afford or do not wish to own a car. Even teens can get Uber accounts now with parental support so they can get rides when needed before they learn to drive.

In the years since the introduction of these ride services, cities have fundamentally changed how they think about parking and curb use. If people want to get a ride and be dropped off, the need for parking and curb use at work, shops, restaurants, and other venues changes dramatically. Airports now use their parking garages more often as a place to get into an Uber or Lyft, than to park a car during your trip.

The rise of ride services such as Uber and Lyft also heightened the concern in cities about the impact of new technologies. Cities have had two key transportation-related complaints about ride services: the increased number of car rides/miles driven and the idea that those rides might take riders from transit. The negative externalities of ride services are the same as those for personally owned cars—more miles and more trips contribute to traffic and pollution. Ride service trips in any city are just a tiny fraction of all car trips. Thus, regulating ride services alone does not change the problem that cars cause in cities.

Regulate All Cars

Cities need to regulate all cars, with effective road pricing, not just regulate or tax ride services. Policies that focus solely on traffic caused by Uber and Lyft are politically motivated, as cities believe that regulating private car use is unpopular, but the unintended consequence of such regulations and taxes on ride services is that people will just drive their own cars. People were not riding transit before Uber and Lyft, and they will not do so now if ride services are penalized and made less attractive. They will instead choose to own and drive their own cars, as they did before. Those who cannot own or drive a car will be disadvantaged once again.

Instead, cities should focus regulations on the effective use and pricing of road space at different times and places during the day for *all cars*—not just ride services. By regulating all car trips now, using road pricing, and reallocating road space to other modes, cities can reduce the incentives for car travel that exist today and prepare for the use of autonomous vehicles in the future.[4]

Cities can implement road pricing mechanisms to deter personal car use in congested areas during peak times. Cars suffer from inverse network effects—the more cars, the worse the experience.[5] This is true whether you are driving your own car, or riding in an Uber or Lyft, or in an autonomous vehicle. Today, our use of cars is not fully priced and many people believe roads are free because we are not charged by the mile, but road usage has massive externalities, such as traffic, pollution, and climate change, which are not paid for with current gasoline taxes. Cities should fully price *all* car rides to reflect the amount of road space they use and the

---

[4] As usual, the future is already available today in San Francisco, with Waymo and Cruise offering fully autonomous ride services with no driver or human in the car other than the rider on board.
[5] Dediu and Bruce, *Micromobility: The First Year* (Asymco, 2019), pp. 210–11.

externalities of car travel.[6] Pricing the road for all cars will reduce congestion and encourage the use of other modes.

Cities should also reallocate space on our streets to make room for non-car modes. Our cities today were built for people to drive and park, at work and shopping areas, restaurants, and events, but parking needs have declined as more people want to be dropped off by a ride service, ride bikes or scooters, or stay home and have a delivery brought to them. Our roads have not changed to keep up with these new ways people want to get around. Cities can rectify this mismatch of supply and demand by opening up more lanes on our streets for uses other than cars. Cities can and should remove street parking in downtown urban areas and use those lanes for through traffic, ride service drop-offs, protected micromobility lanes, and bus-only lanes.

---

[6] "Shared Mobility Principles for Livable Cities, #7," accessed at https://www.sharedmobilityprinciples.org/ ("Every vehicle and mode should pay their fair share for road use, congestion, pollution, and use of curb space.").

2

Price Road Use Fairly

Road pricing? Wait, I thought roads were free. Not exactly, though today we are not being fully charged for the cost of driving. We pay taxes that help pay for the roads and their upkeep, but these taxes do not cover the full cost of driving on our roads. "Driver tax and tolls pay for only about half the cost to build and maintain the physical infrastructure needed to drive. The remainder is paid for with general tax revenues, including 10% from municipal bonds issued to pay for new road projects."[1] There is a growing funding gap because revenue from gasoline taxes is declining due to inflation and reduced use of gasoline with more fuel-efficient, hybrid, and electric vehicles.

The American Society of Civil Engineers estimates that "by 2025, the US will see a $1.1 trillion-dollar shortfall for transportation funding at the federal, state and local levels. Without a long-term solution to fund the nation's transportation system, the Highway Trust Fund will continue to fall far short of meeting our nation's needs."[2] Today, you pay a gas tax every time you fill up your car with gasoline, so the more miles you drive, the more tax you pay. However, more and more cars will be electric going forward and will not require gasoline, but will still cause wear and tear on the road.

A number of cities and states are considering "road user charge systems," which would charge based on miles driven and other costs of road use, such as wear and tear or pollution. In addition,

---

[1] Janette Sadik-Khan and Seth Solomonow, *Streetfight* (Penguin Books, 2016), p. 27.
[2] National League of Cities, *Fixing Funding by the Mile: A Primer and Analysis of Road User Charge Systems*, p. 6.

some cities are considering "congestion pricing" or "cordon" fees, which charge for access into certain congested areas during peak traffic hours. These fees based on peak demand can also be called "surge pricing"—a term popularized by Uber and Lyft to reflect higher prices for times and places where demand for rides exceeds supply. In the future, some combination of these charges will be used to pay for road upkeep. Drivers will likely "pay as you go" for road use, and not all miles will cost the same. In rural areas without much traffic, miles may be cheaper and vary less, but driving into congested areas of cities at peak times will likely incur "surge pricing" for those miles or cordon fees to access downtown areas.

Pay by the Mile

One way to pay for road use is to charge people based on the number of miles they drive on the roads. This is often called a "vehicle miles traveled" or VMT fee. Fees based on VMT use the same principle as the gasoline tax—you pay for how much you use, but VMT fees are charged by the mile instead of the gallon.

This "pay as you go" approach would acknowledge the transition away from gas-powered cars and toward fully electric vehicles, while preserving the idea that those who use roads the most should pay accordingly for road upkeep. The National Association of City Transportation Officials (NACTO) has noted that "VMT fees are assessed based on the number of vehicle miles traveled. By directly pricing travel, VMT fees ensure stable revenue in light of changing vehicle fuel economies and ownership models. Over time, VMT fees could replace gas taxes and help fund infrastructure on a large scale."[3]

VMT fees attempt to offset the "bundle of trips" problem with cars, where you pay upfront to buy a car and then drive it for every possible trip because you have already incurred the majority of the expense of owning it. If driving each mile has additional costs, then the calculus for whether to use your car for every trip will

---
[3] NACTO, *Blueprint for Autonomous Urbanism* (2nd ed. 2019), p. 61.

change. The US Department of Transportation has observed that "Fixed costs of vehicle ownership, such as insurance costs or registration fees, do not currently depend directly on the amount the vehicle is driven. Projects in this category are designed to convert those fixed costs into costs that vary according to the miles the vehicle is driven, thus giving the driver the incentive to recognize these costs when making the decision to drive. Strategies in this category are unique in providing drivers direct financial savings for reducing their driving. Advanced projects relying on Global Positioning System (GPS) may be able to make an even finer distinction for pricing of auto use according to time and location of travel."[4] The State of Utah has announced that it will pilot a comprehensive road usage charge using a "GPS-enabled dongle" operated by ClearRoad. With this sophisticated technology, "governments may soon be able to start thinking about bringing all of their driving fees onto a single bill via a device embedded directly in cars, rather than charging motorists separately at the gas pump, the toll booth, the congestion pricing camera, the DMV and so on."[5] Another version of this pay-as-you-go method could include higher rates charged per mile for larger vehicles that cause more wear and tear on the roadway and lower rates charged per mile if a vehicle carries multiple people or if the driver qualifies for low-income discounts.[6]

Traffic Is Caused by Demand

Why do we have traffic in cities today? Jeffrey Tumlin of the SFMTA (San Francisco Municipal Transportation Agency) explains that "congestion is simply what happens when the demand for mobility equals the supply." For decades, cities have tried to build more roads or more lanes on the freeway, but increasing the supply of roadway has not solved congestion, because the demand for car

---

[4] U.S. Dept. of Transportation, Federal Highway Administration, Office of Operations – Congestion Pricing, accessed at https://ops.fhwa.dot.gov/congestionpricing/strategies/not_involving_tolls/pay_drive.htm.
[5] Kea Wilson, "Utah's Road Usage Charging Pilot Could Finally Price the Roads Properly," *Streetsblog USA*, July 12, 2023.
[6] Ibid.

travel simply rises to meet the available supply. Tumlin notes that "every city in the country has tried to use infrastructure to solve congestion. It has never once worked. *The only way to solve congestion problems is through economic solutions.*"[7] This means we need to use price mechanisms to regulate demand for road space rather than give it away for free.

According to NACTO, the number of vehicle miles traveled in the US alone "hit an all-time high of 3.2 trillion miles in 2017" and "US drivers spent an average of 97 hours sitting in traffic that year." NACTO notes that "Absent policy mechanisms and incentives to encourage people to drive less, traffic will continue to increase. A well-documented behavioral phenomenon, called 'induced demand,' shows that as governments build more or wider roads, people drive more and congestion gradually increases."[8] Induced demand in transportation refers to the principle that building more roads or additional lanes on roads leads to more car traffic to fill the new road space until the same amount of congestion is present. As one CityLab article explained, "When you provide more of something, or provide it for a cheaper price, people are more likely to use it. Rather than thinking of traffic as a liquid, which requires a certain volume of space to pass through at a given rate, induced demand demonstrates that traffic is more like a gas, expanding to fill up all the space it is allowed."[9] One prominent example is the expansion of the 405 freeway in Los Angeles in 2014, which cost more than a billion dollars. Six months after the completion of the new lane, traffic studies showed that travel times on the 405 remained the same as before the expansion. "During the peak drive time from four to seven p.m., travel on the northbound 405 from the 10 to the 101 freeways took thirty-five minutes – one minute longer than the same trip in the previous year."[10] Thus, construction of new roads and more lanes is not able to solve traffic congestion, and economic policies are needed to improve traffic. As Andrew Salzberg explained in an Uber blog, "To tackle this issue

---

[7] Interview of Jeffrey Tumlin, SFMTA, *Smarter Cars Podcast–Ep. 35*, Jan. 2, 2020.
[8] NACTO, *Blueprint for Autonomous Urbanism*, p. 58.
[9] Benjamin Schneider, "CityLab University: Induced Demand," *CityLab*, Sept. 6, 2018.
[10] Sadik-Khan, *Streetfight*, p. 48.

at its core, the cost of driving ultimately needs to reflect its cost to our cities. It's becoming increasingly clear that the most effective way to manage vehicles on the road is through pricing. By charging a fee for all vehicles (private motorists, delivery vehicles, taxis, and services like Uber), road pricing creates an incentive for everyone to share space more efficiently."[11]

Another strategy that has worked is to close or remove roadways rather than widen or increase them. In San Francisco, after the 1989 earthquake destroyed the Central Freeway, the city decided not to rebuild the elevated roadway. "San Francisco's Central Freeway carried around 100,000 passengers per day before it was damaged by the 1989 Loma Prieta Earthquake. The surface-level boulevard that replaced it carries about 45,000 cars. Far from decreasing economic activity, the freeway removal turned the surrounding blocks into one of the city's most desirable (and unaffordable) neighborhoods. Other freeway removals—typically undertaken in dense, central city areas—have been shown to produce similar results. (Bonus: Removing a freeway is often cheaper than repairing it.)"[12]

Traffic congestion "isn't a matter of too little *supply* – roads – it's a product of overabundant *demand* – too many people driving without credible transportation alternatives. Increasing the supply of road space doesn't alleviate traffic; it almost always allows more people to drive more. If building roads actually resulted in less traffic, then surely after 60 years of interstate highway construction we would all be cruising at highway speed. Instead, thousands of road-building and –widening projects have resulted in more lanes, more roads, but no less traffic."[13] Unfortunately, traffic engineers and politicians have been ignoring these outcomes for decades, focusing on increasing the supply of roads rather than addressing the demand for driving in the first place.

With road pricing, cities can use economic solutions to reduce demand by charging by the mile for road usage to encourage

---

[11] Andrew Salzberg, "Road Pricing: A Solution to Gridlock," *Uber Under The Hood – Medium*, Mar. 30, 2017.
[12] Benjamin Schneider, "CityLab University: Induced Demand," *CityLab*, Sept. 6, 2018.
[13] Sadik-Khan, *Streetfight*, p. 62.

transportation choices that fully price in the costs and externalities of driving.

Surge Pricing Will Reduce Traffic

In addition to charging for vehicle miles traveled, cities need to make some miles more expensive to reflect supply and demand to access certain parts of the city at peak times. Some cities have charged these fees as you enter a cordoned area, which is congested during peak times, leading to the name "congestion pricing" or sometimes "cordon pricing"—but the ride service term "surge pricing" is perhaps more descriptive.

Whatever the terminology, using this type of pricing works by encouraging drivers to use alternative travel modes or to travel at different times. "Cordon or zone-based congestion pricing reduces congestion by charging a fee to enter a specific zone or zones of a city...Cordon pricing can be a flat fee or variable, changing over the course of the day to target congestion at peak periods. Cordons can also focus on specific vehicle types (e.g., high polluting vehicles or large trucks)."[14]

Since the main goal of surge pricing or congestion pricing is to encourage fewer drivers during peak times, it is most effective when many drivers will consider other travel modes or times in lieu of paying the fee. As Jeffrey Tumlin, head of the SFMTA, has observed "congestion operates on a 10% margin, so on any street that is gridlocked where none of the vehicles are moving, I don't need to get everyone out of their cars, I just need to convince 10% of those motorists to do something else – travel at a different time, take a different street, or take a more efficient mode of transportation."[15] For congestion pricing to work, cities need to have good alternatives to driving that people can choose when the balance of cost/convenience is changed by the addition of new fees.

The US Department of Transportation explains that "Congestion pricing works by shifting some rush hour highway travel to other

---

[14] NACTO, *Blueprint for Autonomous Urbanism*, p. 60.
[15] Interview of Jeffrey Tumlin, SFMTA, *Smarter Cars Podcast – Ep. 35*, Jan. 2, 2020.

transportation modes or to off-peak periods, taking advantage of the fact that the majority of rush hour drivers on a typical urban highway are not commuters. By removing a fraction (even as small as 5%) of the vehicles from a congested roadway, pricing enables the system to flow much more efficiently, allowing more cars to move through the same physical space."[16] If there is no reasonable alternative to car commuting, no amount of fees will actually reduce traffic in a measurable amount. People will continue to pay higher and higher surge pricing fees because they have no other viable option.

NACTO has observed that "In places where there are fewer alternatives to driving, people's willingness to pay goes up. In 2018, a dynamic toll was enacted along I-66 leading into Washington, D.C. The toll peaked later in the year at $46 due to high demand."[17] If there is little cost to choosing to commute by private car and the transit options are equally long and not much cheaper, it is no wonder a city has congested commutes.

If the transit buses can travel in their own lane, avoid traffic, and provide a reliable and fast alternative to car commuting, then a surge pricing strategy is more likely to work to reduce car use. NACTO urges cities to spend revenue from congestion pricing to improve transit and other modes: "The full benefits of congestion pricing can only be realized when transit service is a viable and attractive alternative to driving. While cities can already take steps to improve transit's frequency and reliability, a commitment to using future revenue to expand transit options is essential to encourage and sustain mode shifts. Cities should also dedicate revenue towards investments in active transportation to support walking and biking."[18] Cities should therefore improve transit and micromobility alternatives to car commuting hand in hand with surge pricing to increase its effectiveness. By using the surge pricing fees collected for further improvements to transit and micromobility options, cities can continue to encourage non-car options.

---

[16] U.S. Dept. of Transportation, Federal Highway Administration, Office of Operations – "What Is Congestion Pricing?" accessed at https://ops.fhwa.dot.gov/congestionpricing/cp_what_is.htm.
[17] NACTO, *Blueprint for Autonomous Urbanism*, p. 63.
[18] NACTO, *Blueprint for Autonomous Urbanism*, p. 59.

Successful City Programs

Several cities have had success in reducing traffic using congestion pricing, most notably London and Stockholm. These are examples that now New York City and Los Angeles are poised to follow. In London, the cordon fee was imposed in 2003 and was immediately successful in reducing congestion. "Even though London experienced a 20% population growth, there was a 9.9% decrease in traffic volume between 2000 and 2015."[19] London also modified its congestion pricing program several times to change the boundary of the cordon and to reduce the exceptions available to the program. Initially, taxis and other ride services were exempt from congestion charges, but with the rise of Uber and Lyft, travel times again increased until the exemptions were removed in April 2019. Similarly, in 2013, the standards for low-emission vehicles that would be exempt from charges changed, and now, there is only an Ultra Low Emission Discount, which set "a threshold that no internal combustion engine could meet" in 2013. Transport for London expanded the Ultra Low Emission Zone across all London boroughs on August 29, 2023.[20] These adjustments allowed London to continue to reduce car traffic and air pollution. "London's policy has proved a success: Since it was introduced in 2003, the number of private vehicles entering central London has declined by 39%. If a city makes driving more expensive, it must be ready with alternative and affordable transit options… Congestion pricing, when done right, should improve bus speed by moving a good chunk of car traffic out of their way, and the city should waste no time in proving that logic out."[21]

Stockholm also had similar success with congestion pricing, introduced in 2007. Average traffic volumes into the cordoned area during peak hours dropped by 22%.[22] In Stockholm, "the biggest evolution has been the changing of public attitudes in favor of the

---

[19] National League of Cities, *Making Space: Congestion Pricing in Cities*, p. 6.
[20] Transport for London, "*Ultra Low Emissions Zone,*" accessed at https://tfl.gov.uk/modes/driving/ultra-low-emission-zone/ulez-expansion-2023.
[21] Eillie Anzilotti, "5 Lessons for New York's Congestion Pricing from Cities that Have Made It Work," *Fast Company*, April 2, 2019.
[22] National League of Cities, *Making Space: Congestion Pricing in Cities*, p. 8.

congestion tax...This shift in perception is largely due to the fact that revenues have been funneled into road improvements outside of the inner city, allowing residents throughout the area to enjoy the benefits."[23] If fees from congestion pricing can be used to improve commute options throughout the area, including public transit systems that can be used as an alternative to driving, then the system can become more equitable and enjoy broad support.

When New York implements congestion pricing, now scheduled for April 2024, the fees are estimated to run as high as $23 per car and those funds will be used to improve the subway and other public transit systems that are in need of upgrades and repairs. The plan is projected to raise up to $1 billion a year for public transport projects, and officials estimate "the charge could reduce daily traffic in the district by 20%."[24] New York's plan has taken years to clear regulatory approvals and still faces litigation challenges from the state of New Jersey.

Even Los Angeles, the most car-dependent city in America, is now considering congestion pricing to address its infamous levels of traffic. Los Angeles City and transit officials began conducting feasibility studies in 2019 to examine the rollout of a congestion pricing pilot project. The latest estimates suggest that it may take until 2028 to implement a pilot.[25] It remains to be seen how New York and Los Angeles ultimately implement these surge pricing fees, but the fact that the two biggest cities in the US are finally coming around to the idea shows a mind shift in the right direction.

One barrier to implementing new congestion pricing systems has been the cost to implement cordons to charge for access. Given the upfront cost to implement the system over the last few decades, it has been difficult to adopt pilot programs to test out how a system might work in practice. As Robin Chase explains, "the challenge has been, how do you do a congestion pricing pilot when it's going

---

[23] National League of Cities, *Making Space: Congestion Pricing in Cities*, p. 9.
[24] Michelle Kaske, "What Congestion Pricing's Arrival in NYC Would Mean," *Bloomberg CityLab*, May 12, 2023.
[25] L.A. Metro, "Traffic Reduction Study," accessed at https://www.metro.net/projects/trafficreduction/ on Aug. 27, 2023 (forecasted opening of 2028).

to cost you $350 million to build the infrastructure?"[26] This is another area where technology improvements can change the political calculus and pilots can be implemented using a cell phone, FasTrak, E-Z Pass, and other lightweight technologies.

Political Changes

New technologies and new attitudes are also changing the political calculus around congestion pricing. As NACTO has acknowledged, "Until recently in the US, elected officials have been reluctant to embrace meaningful congestion pricing. However, perhaps due to the rise of ride-hail services, the way that people think about transportation payment and pricing is evolving. Consumers are quickly becoming accustomed to 'surge,' variable, and peak pricing. Electronic tolling systems and payment platforms have made it more convenient to pay for travel."[27] Just as residents are getting used to Uber and Lyft ways of pricing rides, including surge pricing at peak times, they are also beginning to think of car rides as a service.

Policy support is also coming from generational change. Young people are not as wed to owning and driving cars. If you do not own a car, the whole idea of driving becomes less important and just one of several transportation choices you can make. Today, it seems that fewer young people value driving and the number of 16-year-olds obtaining drivers' licenses has declined from 46.2% in 1983 to just 24.5% in 2014, and by age 19, only 69% had licenses in 2014, compared with 87.3% in 1983.[28] As a Wired article noted, "Another reason congestion pricing suddenly feels possible: Today's city residents, especially young ones, seem to see cars more as transportation tools than as extensions of their identity. Which means they might be more willing to give up that steering wheel for alternatives, like transit, bikes, or ride-hail apps."[29] This is part of a broader shift in society from valuing the car as an asset and

---

[26] Interview of Robin Chase, *Smarter Cars Podcast – Ep. 29*, Oct. 22, 2019.
[27] NACTO, *Blueprint for Autonomous Urbanism*, p. 58.
[28] Julie Beck, "The Decline of the Driver's License," *The Atlantic*, Jan. 22, 2016.
[29] Aarian Marshall, "The Age of Congestion Pricing May Finally Be Upon Us," *Wired*, Feb. 28, 2019.

reflection of identity to viewing car rides as simply another form of transportation.

Finally, many people believe that using surge pricing to solve congestion is unfair to lower-income drivers because it is a flat tax that rich people can more easily afford. However, this policy concern ignores the broader picture of how the need to own and drive a car, and sit in traffic, is itself an expensive burden on lower-income families. As Jeffrey Tumlin has noted, "right now the system is structured to grossly subsidize those who are the wealthiest and to penalize the people who are making the most efficient use out of the mobility system."[30] If the fees from congestion pricing are used to improve transit options, then the burden of driving can be removed and time saved for everyone. For example, in New York, you can drive into Manhattan for free over at least one bridge while taking the subway in costs $6. So, those who can afford cars are receiving a subsidy, while lower-income residents riding the subway pay more. Congestion pricing and VMT taxes can generate funds to be used to improve and subsidize transit, benefiting low-income residents who rely on transit. Cities can also provide income-based discounts for car drivers who need assistance to offset road-based pricing.

---

[30] Interview of Jeffrey Tumlin, *No Parking Podcast – Ep. 18*, Feb. 23, 2020.

Reallocate Space on Our Streets

Today, our cities are designed for cars not people—from our street design to our allocation of land, we have carved out space to drive and park automobiles wherever we go. There are few places in cities where people can walk, bike, or ride a scooter safely, or stroll to find a shop or dine outside. We dedicate a huge portion of our land to roads, parking lots, garages, and curb spaces. Jane Jacobs wrote eloquently about this "erosion of cities" by automobiles. She described a process that happened slowly as a result of small decisions "nibbling" away at the city.[1] Likewise, she suggested a similar approach to the "attrition" of automobiles—where cities could make small decisions one by one to make car use in cities less attractive and make cities more inviting for all. We find ourselves some decades later needing more than just small changes to take back the erosion of cities by automobiles. To even the playing field, we must reallocate to other modes some of the space now dedicated to cars.

In the last decade, we have gained a powerful new tool in this fight to reclaim our cities. Thanks to new mobility technology, we finally have attractive alternatives to personally owned cars. New mobility options such as ride-hailing and micromobility are incredibly popular, but these options have also created new challenges for cities that cannot be solved by technology alone. We need to combine these new mobility modes with urbanist policies to keep our roads moving, in order to start the "attrition" of automobiles in cities that Jane Jacobs envisioned.

---

[1] Jane Jacobs, *The Death and Life of Great American Cities* (Vintage Books, 1961), p. 349.

Cities have become hostage to a single mode—cars—making every other mode less attractive and often unsafe. This is the ultimate chicken and egg problem for transportation planners. When there are few safe bike lanes, fewer people ride bikes and scooters, and cities feel uneasy about allocating space that may go unused. As Robin Chase explains, "we make our choices based on what's convenient and what's economical, and we've spent the last hundred plus years making personal cars convenient and economical and everything else harder."[2] To restore a balance among modes, allowing multimodal transportation to thrive, cities need to reallocate existing space on roads to make room for micromobility and transit, and to accommodate more productive uses for curb space.

Parking Is for Garages

Most cities in the US were built, or rebuilt in the last few decades, to accommodate people driving cars and parking them everywhere they go. On most streets, two whole lanes are reserved for car parking, one on each side of the street.

Cities are changing, with new technologies providing new ways to get around. The availability of ride services has encouraged many city dwellers to leave their cars at home and get a ride to work, dinner, or events in an Uber or Lyft. Cities point to the large number of ride service trips as a source of traffic. Many fewer people are driving themselves into downtown business districts, where revenue at garages and meters has declined. This shift away from parking and toward ride services shows that cities are in transition.

We are moving from a system that was primarily "drive and park" to one where more people prefer to "ride and drop-off." Yet, our allocation of street space has not kept up with this shift. We have not made a matching transition of land use from street parking to travel lanes and drop-off curb zones. By shifting space

---

[2] Interview of Robin Chase, *Smarter Cars Podcast – Ep. 29*, Oct. 22, 2019.

from "drive and park" uses to "ride and drop-off," we can make room on our streets for this transition.

One way to improve traffic flow in downtown areas is to remove street parking wherever possible. Cities need every lane on the street in congested downtown areas to be used for moving people and vehicles, not for storing cars. Unless there is no garage or lot nearby, which is a very small number of places, all street parking downtown should be removed. Why should people store their cars on a public street and take up space that could be used more productively? Cities should use these lanes for travel and for pick-up and drop-off lanes for ride services and delivery vehicles, which might otherwise block traffic by double parking when there is no space at the curb to pull over. Traffic in downtown areas is often caused by people circling around looking for street parking. Studies in 15 cities dating back decades show that "cruising for parking accounted for between 8 and 74% of traffic in the areas studied and the average time to find a curb space ranged between 3.5 and 14 minutes."[3] By removing street parking and directing people into garages or lots, that traffic can be eliminated. By repurposing the curb from street parking that stores cars to pick up/drop off and delivery zones, cities can improve traffic flow and gain additional lanes for traffic.

Protected Lanes for Micromobility

Cities can also reallocate road space to make room for micromobility. As discussed in Part II, people on bikes and scooters need networks of micromobility lanes, preferably protected lanes, to ride more safely. Protected lanes have a physical barrier between the lane and the car traffic. These barriers can be plastic sticks coming out of the ground or other creative barriers, or a row of parked cars moved out from the curb enough to create a lane between the cars and the sidewalk. Both types of protection are inexpensive and easy to install.

---

[3] Donald Shoup, *Parking and the City* (Routledge Press, 2018), p. 22.

Cities can make space for micromobility lanes by removing street parking or by making existing lanes narrower. On a typical street with two lanes of parking and two lanes of traffic, each lane is allotted 12 ft across. As Janette Sadik-Khan explains, "Reducing the width of the two parking lanes that flank the street from twelve to just nine feet leaves more than enough room to park even an oversized vehicle…this simple change can yield six full feet of space that can now be reprogrammed for other uses."[4] Many cities are creating mobility lanes using the row of parked cars on the street as a protective barrier, by making the mobility lane run along the sidewalk and floating the parking lane for cars away from the sidewalk. "If we place the bike lane where the parking lane was, the parking lane becomes a 'floating' lane, parallel to but not alongside the curb…[i]f we narrow the two moving lanes from twelve to ten feet each, there is enough room for moving traffic and an additional four feet of roadbed. This reclaimed space can be added as a buffer between the bike lane and the parking lane so the car doors of people getting out of their vehicles don't swing into the bike lane and 'door' a passing bike rider."[5]

Cities should use these street design techniques to reallocate space to create a network of protected lanes from downtown/business areas to and from transit centers, neighborhoods, and major corridors where commuters ride. This system of protected lanes only works if it provides a complete solution for commuters. Therefore, putting in a block here and there, over the course of three to five years, does not move the needle to encourage people to get out of their cars. The best way to change habits and get people out of cars is to create an entire system of protected lanes that will get people where they need to go.

Protected lanes are also good for car traffic. The best way to make traffic move faster is to get people out of cars who are traveling shorter distances and who do not otherwise need to ride in a car due to disability, age, cargo, fancy dress, or other reasons. Those who are willing and able can use smaller electric vehicles—such as electric bikes and scooters—for these shorter trips in cities.

---

[4] Janette Sadik-Khan and Seth Solomonow, *Streetfight* (Penguin Books, 2016), p. 51.
[5] Sadik-Khan, *Streetfight*, pp. 53–54.

Micromobility does not work for every person or every trip, but for those it does, we should make room for them to ride more safely. While many people are interested in riding electric bikes and scooters, many more would do so if they had a protected place to ride. By reallocating space on our streets, cities can increase the ridership of micromobility vehicles.

Let Buses Run Free

Similarly, cities can make room for bus lanes by removing downtown street parking or reducing the width of lanes. As discussed in Part III, buses can run faster and on time when they have their own bus lanes. This mostly requires signs and/or red paint, which are not expensive. Drivers may think giving buses a lane is bad for cars, but it actually gets the buses out of the car lanes where they tend to block traffic each time they stop. If we take away street parking in the congested downtown areas, we are adding a lane that buses can use without taking away car lanes at all. If buses run faster and on time, then more people who can ride them will do so. This will reduce car traffic in the remaining lanes.

By taking these steps to reallocate space on streets, we can save lives, improve safety, move more people, and reduce traffic. Even if just as pilot projects, cities should move forward to implement these changes quickly and iterate over time as needed.

Policy Change Is Possible

Going back to Jane Jacobs in the 1960s, urbanists have advocated for better ways to make city streets work for pedestrians, cyclists, and buses—not just cars, but efforts to change policy to implement bike lanes, bus lanes, and congestion pricing and other efforts to reshape our streets have been defeated by car-centric voters and policymakers. For example, cycling and pedestrian advocacy groups in San Francisco and other cities have advocated tirelessly for better infrastructure and protected lanes forever. Despite their heroic efforts, over decades, success has been limited and slow to happen, with projects often taking three to five years or longer. In New York, it was a huge fight to add protected bike lanes and bike share systems. Janette Sadik-Khan wrote a whole book about it called *Streetfight*.[6] For decades, car owners insisted that city officials continue policies that would reserve street space for cars rather than make protected lanes or take other measures to enable safe cycling.

Then, in 2018 and 2019, the political tide seemed to shift in favor of carving out space for people on our right of ways. Protected lanes were expanded and implemented with more urgency, and advocacy seemed to have greater results.[7] What happened? The introduction of ride services and micromobility improved the ability of cities to implement new transportation policies. In particular, a new technology came along that changed the equation. The arrival of micromobility vehicles changed the political dynamic.[8] In 2018, dockless electric scooters arrived on the scene, and the use of electric bikes increased, together adding many new riders pushing for safe street spaces. Robin Chase, new mobility advocate and founder of Zipcar, observed that electric scooters are bringing in "a whole new group of users" who have never ridden a bike or been in a bike lane before.[9] Suddenly, the allocation of space on the street for small electric vehicles has become part of a larger movement that affects more people than just hard-core cyclists.

People who would never ride a regular bike can imagine using an electric scooter or an electric bike to get to work. The Mayor of San Francisco, London Breed, noted her support for the expansion of the electric scooter program in San Francisco by saying: "I like scooters, you can ride 'em in a dress."[10] Electric bikes also require very little effort and allow people to commute without needing a shower when they arrive at the office. Electric bikes and scooters give people a vision of a commute without a car that can still

---

[6] Janette Sadik-Khan and Seth Solomonow, *Streetfight* (Penguin Books, 2016).
[7] See the "quick build" projects adopted by SFMTA, accessed at https://www.sfmta.com/vision-zero-quick-build-projects.
[8] Interview of Sanjay Dastoor, CEO of Skip, *The Micromobility Podcast - Ep. 58*, Feb. 5, 2020 (noting political environment changed for bike lane infrastructure after introduction of electric scooters).
[9] Interview of Robin Chase, *Smarter Cars Podcast - Ep. 29*, Oct. 22, 2019.
[10] Joe Fitzgerald Rodriguez, "Mayor Breed Signals Support for SFMTA Scooter Policy," *SF Examiner*, Sept. 29, 2019.

be convenient, reliable, and enjoyable. This vision has created a whole new wave of city dwellers advocating to reallocate street space away from cars to make room for micromobility.

Another example is street parking. With the rise of ride services, delivery services, and e-commerce, there is now more political support to remove street parking altogether in downtown areas. Instead, cities want to use the curb space for pick-up and drop-off zones to allow ride services and delivery vehicles to pull over and not block traffic. Cities are starting to realize that they already have too much parking just with downtown garages, as people start to use ride services and micromobility to get to work. Jeffrey Tumlin, head of the SFMTA, has noted, "We're seeing year over year pretty steady decline in parking demand, even as economic activity increases and population and employment increase in San Francisco, and so in order to keep our off-street garages full, which are increasingly empty, it's actually beneficial to ratchet down some of the on-street supply in places where it is non-essential."[11] Those street lanes could be used more productively for through traffic, for pick-up and drop-off, and for protected lanes for bikes and scooters. This has become even more true in the years since the pandemic as fewer people work in offices downtown, and garages and office buildings sit empty in many cities.

Together, new mobility advocates and city planners can bring new ideas and new voters to support changes to our city infrastructure to make space on our streets for transit and micromobility, not just cars. By pricing our streets fairly and allocating more space to non-car modes, cities can take advantage of new mobility technologies to encourage multimodal transportation and change how we use cars in cities.

---

[11] Interview of Jeffrey Tumlin, SFMTA, *Smarter Cars Podcast – Ep. 35*, Jan. 2, 2020.

Part II
Make Micromobility Work

Micromobility should be defined around what is singular about its purpose: moving a human being...Its purpose is thus to offer maximum freedom of mobility and its minimalism is to do so in the least impactful way. Its minimalism means it needs to leave no trace of itself and ask the least for itself.[1]

—*Horace Dediu*

---

[1] Horace Dediu, "The Micromobility Definition," *Micromobility Industries*, Feb. 23, 2019.

4

Invest in Infrastructure

Micromobility Is Good for Cities

Micromobility is the perfect marriage of technology and urbanism: It offers the convenience and utility of a car, with a lower carbon footprint, the joy of a bike ride, and the road geometry of public transit. It is the only mode of travel that matches the three benefits of driving a car—point-to-point travel, with no fixed schedule, in a private space—while also providing good road geometry. As cities face more traffic and competition for road space, micromobility can help replace shorter car trips with smaller modes that will improve traffic, land use, pollution, and well-being in cities. As 50% of all car trips are three miles or less, micromobility is well suited to replace many of these trips.[1]

Micromobility and transit are both good for cities and together can serve more people with attractive options. Why should cities not build space on the public right of way for micromobility like public transit? In the same way that cities provide bus lanes and bus stops for public transit, cities should also invest in parking and riding infrastructure for shared micromobility programs. A narrow focus on public transit has caused cities to undervalue and over-regulate shared micromobility, rather than giving it the necessary space it needs on city streets.

The prioritization of transit above all other modes is a misinterpretation of cities' own "Transit First" policies. In fact, these

---

[1] Darren Flusche, "National Household Travel Survey – Short Trips Analysis," *League of American Bicyclists*," Jan. 22, 2010.

policies were designed to promote all modes that are not personally owned cars and "Transit First" is a misnomer. For example, the actual text of the Transit First policy in San Francisco, first adopted in 1973, calls for the promotion of walking and biking as well as transit, and even includes taxis as part of its definition of public transit:

> Public transit, including taxis and vanpools, is an economically and environmentally sound alternative to transportation by individual automobiles. Within San Francisco, **travel by public transit, by bicycle and on foot** must be an attractive alternative to travel by private automobile.
>
> San Francisco Charter, Sec. 8A.115 (emphasis added)

The policy specifically references the need to allocate street space for these modes with good road geometry:

> Decisions regarding the use of limited public street and sidewalk space shall encourage the use of public rights of way **by pedestrians, bicyclists, and public transit,** and shall strive to reduce traffic and improve public health and safety... **Bicycling shall be promoted by encouraging safe streets for riding, convenient access to transit, bicycle lanes, and secure bicycle parking.**
>
> San Francisco Charter, Sec. 8A.115 (emphasis added)

Policies like these in cities were designed to shift travel modes from private cars to this group of beneficial alternatives. They were not designed to prioritize transit over micromobility—the use of which city planners at the time would have viewed as a huge benefit not a problem. Micromobility should not be required to prove that its riders only replace car trips and not public transit use or walking to be a valuable addition to cities. Data from cities that have implemented micromobility programs show that riders often use these services for a combination of trip purposes, including replacing short car trips, enhancing first-mile/last-mile connections to public transit, and substituting for walking or cycling in certain situations. These all promote multimodal travel. The North American Bike Share Association (NABSA) reports in 2022 that in North America 37% of shared micromobility trips replace car trips and 64% of riders reported using shared micromobility to access transit.[2] Scooters are often accused of simply "disrupting walking," but even that use case is valid, for example, where a woman is traveling alone at night and finds a scooter safer than walking, or someone wearing a suit and dress shoes uses a scooter to ride longer than would otherwise be comfortable to walk. Offering people a combination of different modes and options, and not just public transit, increases the likelihood they will stop relying on private cars as the default mode.

Cities should be agnostic between modes with good road geometry. Buses achieve good road geometry by grouping many people together on one big vehicle, but micromobility achieves the same road geometry as each person rides her own bike or scooter in the bike lane. In each case, the rider takes up a tiny amount of space on the street, allowing for more people to get around without clogging traffic. Whether cars are electric, autonomous, or otherwise, they still have terrible road geometry, taking up about 300 ft² of road space while mostly moving just one person. In contrast, riding scooters, bikes, or the bus takes less than 2 ft² person.[3] In San Francisco, Jeffrey Tumlin of the SFMTA explains that "Geometry is the simplest and most profound driver of decision-making for a transportation agency – we are not getting streets that are any wider and we've got to move more people, so that means moving more people in the same amount of space and

---

[2] North American Bike Share Association, "2022 Shared Micromobility State of the Industry Report," Aug. 10, 2023.
[3] Interview of Jeffrey Tumlin, SFMTA, *Smarter Cars Podcast – Ep. 35*, Jan. 2, 2020.

prioritizing the modes of transportation that are most space-efficient."[4] At the rate of one person per car, cities cannot move people in cars without massive traffic jams and pollution, but micromobility can move more people in smaller bike lanes. By adding this option alongside public transit, cities can encourage more people to move in space-efficient ways.

Micromobility fleets in cities make electric bikes and scooters so easy to use that millions try them for the first time and discover a new way to move through cities without a car. In 2022, micromobility trips rebounded to pre-pandemic levels with 157 million trips on 289,000 vehicles across 401 cities just in North America.[5] Globally, McKinsey has reported that micromobility trips were 16% of all trips in 2022, even with the limitations currently facing micromobility riders.[6] Lime, one of the largest micromobility operators globally, reported a record first half of 2023 numbers, including more than 40 million trips taken in Q2 2023 alone.[7] Micromobility is popular and offers an attractive alternative to cars, but it requires space on the public right of way. As Haya Douidri, EVP of Policy and Strategy at Superpedestrian, observes, "While current micromobility policies may primarily focus on risk mitigation, such as avoiding sidewalk clutter or traffic conflicts, it is equally essential that cities seize the full potential of bikes and scooters and create an environment that maximizes their usage to encourage a shift away from private cars." Not everyone will give up a car trip for a bus, but micromobility provides another attractive alternative to cars and can serve first- and last-mile trips to transit as well. By providing public infrastructures such as parking spots and bike lanes, cities can promote micromobility as an important non-car mode of transport.

---

[4] *Ibid.*
[5] North American Bike Share Association, "2022 Shared Micromobility State of the Industry Report," Aug. 10, 2023.
[6] Kersten Heineke, Nicholas Laverty, Timo Möller, and Felix Ziegler, "The Future of Mobility," *McKinsey Quarterly*, Apr. 23, p. 2.
[7] Alyssa Harris, "Lime Breaks Its Own Record with $27m Netted in H1," *ZagDaily*, Sept. 13, 2023.

Micromobility Needs Infrastructure

As more and more people ride electric bikes and scooters, cities need to build the infrastructure for riding and parking these vehicles. "If cities want to promote scooters as car-alternatives… they should evaluate how to make scooter travel safe and comfortable. Part of this effort should consider how cars can impede access for both scooters and bikes…Cities wishing to promote space-efficient modes like bikes and scooters should therefore enforce not only how micromobility vehicles can block pedestrian travel – the focus of most scooter parking requirements – but also how cars can obstruct bike, scooter, and pedestrian travel by obstructing bike lanes or parking across sidewalks."[8] Each mode needs a safe place to operate that does not bring the driver, rider, or walker in conflict with other modes.

In truth, most cities, especially in the US, simply never had many bike or scooter riders on their streets until shared fleets came along and therefore do not have the parking or bike lane infrastructure needed to avoid conflicts with pedestrians, cars, and other modes. Thus, the very success of shared micromobility operators in achieving what transportation planners had long sought—getting more people to use bikes and small vehicles—has led to an ironic backlash as cities try to regulate the conflicts between micromobility and other modes. Cities have struggled to manage the vision of micromobility, which they know can reduce traffic and pollution, with the reality of shared micromobility fleets on the public right of way.

With transportation budgets cut and public transit suffering, cities may not have the money to subsidize micromobility, but cities can do two things: (1) invest in low-cost parking and bike lane infrastructure and (2) reduce unnecessary fees and burdensome requirements that make city permit programs unsustainable for operators, as we will address in Chapter 5.

---

[8] Anne Brown, "Micromobility, Macro Goals: Aligning Scooter Parking Policy with Broader City Objectives," *Transportation Research Interdisciplinary Perspectives*, Volume 12, 2021, 100508, ISSN 2590-1982, https://doi.org/10.1016/j.trip.2021.100508.

A Place to Park

Cities can and should promote micromobility programs with infrastructure that improves operations, safety, and efficiency. Micromobility users need a place to park and a place to ride—cities can build infrastructure to do both.

Giving bikes and scooters a place to park and setting norms for acceptable parking procedures are important to addressing the "clutter" concern, especially about shared scooters. Cities and operators need to find the right balance in different neighborhoods to offer attractive places to park scooters and bikes, without hindering the usability of the system.

Neighborhood rules. The major shared micromobility operators have advocated that cities take a two-pronged approach to parking.[9] First, mandatory parking spots should be used only in dense urban downtown areas, with a minimum of 40 bays and three parking spots per scooter permitted in the city. For specific highly pedestrianized areas, where parked scooters will be a nuisance, cities can require geofencing to enforce no parking zones. Second, the remainder of city areas should be left as free-floating dockless parking, with no mandated parking spots, as parking infrastructure (such as racks and painted areas) is not as readily available and the density/clutter is less likely to cause issues for pedestrians. The operators have suggested "setting parking requirements based on population and local activity, existing infrastructure, and pedestrian traffic to enable a mix of dockless and mandatory parking that meets local needs."[10]

Because of their smaller size, and the fact that they tip over easily, shared scooters have proven more challenging than free-floating bike share to avoid clutter on sidewalks. "Scooters left blocking sidewalks have been a big source of complaints almost everywhere they've been introduced, and cities have evolved

---

[9] "Micromobility Industry Best Practice" Statement by Dott, Lime, Superpedestrian, Tier, Voi and Bird in Europe, December 2022; "Joint Statement of Bird, Lime, Spin and Superpedestrian," in US market May 2023.
[10] Ibid.

different solutions for it. In San Francisco, every scooter must be locked to something, a regulation which has led to the companies innovating new locking mechanisms. Sacramento has created parking zones by painting them on the ground..."[11] The lock-to requirement in San Francisco, Washington, DC, Berkeley, and other cities has been very effective for controlling scooter clutter and has also helped the industry avoid theft, abuse, and vandalism of scooters. Similarly, painting a scooter icon on the sidewalk inside a rectangular border has indicated to riders some acceptable places to park scooters. Both lock-to requirements and visible scooter parking locations on streets help scooter riders understand where scooters should be parked and shift the rider mindset to park scooters as they park personal bikes—locked to a parking rack or other fixture, or parked in a designated parking area.

Many cities also see the benefit of using car parking spots on the street for micromobility parking. By removing a parking spot for cars at an intersection, cities get the benefit of "daylighting" that allows pedestrians to be more visible at intersections. These removed parking spots can also be used for scooter and bike parking with small bike racks, or for free-standing scooter or bike parking. NACTO's Guide to Shared Micromobility Permitting, Process and Participation notes the benefit of using space on streets for micromobility parking: "The street, where people are already riding, is the best place for shared micromobility pick up and drop off. Many neighborhoods that have incomplete or narrow sidewalks and no furniture zones may also have abundant street parking for motor vehicles. To support dockless shared micromobility in these areas, cities should consider all options, including: allowing lock-to at light poles and street signs; allowing dockless devices in striped bulb-outs; and investing in designated on-street corrals. Consider marking on-street corrals in metered motor vehicle parking space or allowing operators and users to leave devices in unmetered street-parking spaces."[12]

---

[11] Melanie Curry, "What Regulatory Role Should the State Play with Scooters, Bike-share? CA Legislators Want to Know," *Streetsblog Cal*, Nov. 6, 2019.
[12] NACTO, "Shared Micromobility Permitting, Process and Participation," Urban Bikeway Design Guide, Dec. 2022, p. 13.

Cities can also encourage more multimodal trips with transit and scooters by building parking racks and designated parking areas near transit stops and stations. The time operators spend rebalancing scooters to those areas can help serve transit, while also generating more trips per day for scooter operators. This infrastructure is a win-win for cities and operators.

To Dock or Not to Dock. Some have suggested that cities should move toward a fully docked scooter model, requiring riders to leave scooters in specified docked areas, similar to station-based bikeshare programs. However, the ability to leave a scooter directly at your destination and not have to look for many blocks for an appropriate dock with space is one of the main benefits of scooter riding. Therefore, the use of mandatory docks or parking spots should only be implemented in the downtown core area where density and pedestrian traffic require more tightly controlled parking. If cities want to encourage a mode shift away from cars, which are entirely dockless and convenient, they need to allow shared scooters to provide truly point-to-point service without the limits of finding a dock.

Dockless bikes and scooters have been much more popular than docked bikeshare in part due to the flexibility to leave the vehicle at your end destination without worrying whether a dock or parking spot will be available. Even Lyft that operates docked bikeshare in cities such as San Francisco has implemented dockless bikes alongside its docked bikeshare programs in some programs to meet the need for more flexible pick-up and drop-off points. As Lime has noted, "Scooter ridership surpassed traditional bike share because of proximity and convenience. In contrast to a docked bike, a 'free-floating' scooter mere steps away raises the odds that someone will choose it instead of a car...Lime's rider-analysis data shows, however, that with each added step that a user has to walk to find a scooter, the less likely riders will choose a scooter as an alternative to a taxi. Instead, a better approach is to have scooters parked no more than 60 steps (150 feet) from any building, available through either free-floating parking in less dense areas or abundant parking spaces in dense downtowns."[13]

---

[13] Adam Kovacevich, Head of Government Relations (Americas) for Lime, "Op-Ed: How Cities Should Regulate Scooters," *Streetsblog NYC*, Oct. 29, 2019.

The Benefits of Local Charging. As cities add mandatory parking areas in dense downtown areas, there is one other option to consider for the future: charging infrastructure for bikes and scooters. To improve operations for shared services and private vehicles alike, cities could add docks that actually charge these electric vehicles and eliminate some of the costs involved with keeping fleets charged. With docks that charge bikes and scooters, operators can incentivize riders to return the vehicle to a dock at the end of the ride in order to earn free credits or other benefits. By eliminating the need for operators to pick up scooters and charge them, or swap out batteries that are charged, cities can reduce the costs of using micromobility vehicles and make the programs more sustainable for operators—all while reducing clutter by keeping scooters in docks for charging when not in use. Swiftmile, Kuhmute, and Metro Mobility Charge Lock are some of the companies developing such docks as charging stations, which can offer more than just a place to tidy clutter. By incentivizing riders to return a scooter to the dock if they can, to earn credits or other goodies, companies can save costs on manpower needed to collect and charge scooters at night.[14] Cities should partner with scooter operators to create charging and parking infrastructure to improve micromobility operations.

A Place to Ride

Cities should also continue to build and expand infrastructure to give bikes and scooters a safe place to ride. By building micromobility lanes, and especially protected lanes, cities can largely solve the issues around scooter riding on sidewalks and potential dangers to pedestrians, while also reducing the other major safety issue for scooters and bikes: crashes with cars. Scooter riders largely ride on the sidewalk to avoid the dangers of colliding with cars on

---

[14] Interview of Colin Roche, CEO of Swiftmile, *Smarter Cars Podcast – Ep. 36*, Jan. 14, 2020.

unsafe roadways that have not carved out space for micromobility vehicles. When bike lanes are available, not only do more people ride bikes and scooters, but they also avoid riding on sidewalks and use the protected space given to them.

In San Francisco, Jeffrey Tumlin of the SFMTA explains that "If we want bikes and other forms of micromobility to actually be well used and to take advantage of their potential...one of the most important elements is a safe, protected place to ride. That also gets all the bikes and scooters off the sidewalk as well...San Francisco has been investing heavily in protected bike lanes, which are actually protected micromobility lanes for scooters, skateboards, bikes...and all of those small wheeled devices."[15] The city of Denver, Colorado announced in 2018 that it would add 125 miles of new bike lanes in five years, and Denver Mayor Michael Hancock exceeded that goal: "In the last five years, we've transformed our streets with new neighborhood bikeways, protected bike lanes and traffic calming measures to slow cars and create a safer citywide transportation network for all," Hancock noted in a statement announcing the city had reached 137 miles of bike lanes in May 2023. Hancock noted the importance of infrastructure to micromobility: "We have set a strong foundation to ensure the future of mobility in Denver includes the ability to travel more easily by bike and offers safe places to scooter as well." Denver now has "more than 433 total miles of bikeways, including on-street and off-street facilities (such as trails)" and that includes "more than 300 miles of bikeways" that are on-street, demonstrating "a citywide commitment to prioritizing streets for people and providing safer connections to eliminate fatal and serious injury crashes."[16]

Some cities have responded to reports of injuries by declaring scooters dangerous and imposing more regulations on operators, but scooters are not the problem, as most serious incidents are collisions with cars on shared roadways. As shared scooter companies have noted, they can improve the scooter vehicles to make them sturdier and improve rider education to encourage people to

---

[15] Interview of Jeffrey Tumlin, SFMTA, *Smarter Cars Podcast – Ep. 35*, Jan. 2, 2020.
[16] City of Denver, "Denver Hits Major Milestone in Bike Infrastructure - City and County of Denver (denvergov.org)," May 23, 2023.

use the vehicles properly, but these efforts can only improve safety around the margins. The real safety improvements will come from infrastructure changes in cities that separate scooter and bike riders from cars. Shared micromobility operators have issued recommendations to cities advocating for building more protected micromobility lanes to improve safety, rather than focusing on other remedies that are less effective, such as mandatory helmet laws.[17]

Protected lanes are a simple fix that cities can implement to improve safety for bikes and scooters, as well as for cars themselves. "A comprehensive study of crash and street design data from 12 cities finds that roads with protected bike lanes make both cycling and driving safer…Separate and protected bike lanes were the strongest indicator of lower fatality and injury rates. Where cycle tracks were most abundant on a citywide basis, fatal crash rates dropped by 44% compared to the average city, and injury rates were halved. While cyclists benefited from having painted bike lanes as well as fully separated bike lanes in terms of safety, what paid off the most for all road users—drivers included—were protected lanes fortified with stanchions, planters, and the like."[18] These infrastructure improvements to add protected lanes are the main change cities can make to improve safety for micromobility riders.

Companies cannot change the environment on public streets and sidewalks to accommodate scooter riding—but cities can. This is why it is so important for cities to make these improvements to encourage micromobility use. These projects are not very expensive compared with traditional infrastructure undertakings. As the founders of micromobility operator Spin noted, "Dropping the word 'infrastructure' generally brings to mind projects that cost billions, or trillions – the kinds of projects that are dropping like flies as city and state budgets evaporate due to the current economic crisis. But for surprisingly little money, there are changes we can make to our streets now that will physically alter our communities

---

[17] "Micromobility Industry Best Practice" Statement by Dott, Lime, Superpedestrian, Tier, Voi and Bird in Europe, December 2022; "Joint Statement of Bird, Lime, Spin and Superpedestrian," in US market May 2023.

[18] Laura Bliss, "Protected Bike Lanes Are Safer for Drivers, Too," *City Lab*, June 3, 2019.

in ways that will make them healthier, cleaner and safer for generations to come. These changes could lure back those who have returned to the automobile, providing them a safe, economical – enjoyable, even – socially distanced alternative."[19] Micromobility operators and advocates for cyclists and safe streets have continued to ask cities to implement infrastructure that will allow more people to ride bikes and scooters safely.

People often point to China and Europe to prove that micromobility and biking can be popular modes in everyday transportation, not just recreation, but in those cities, the government has built substantial infrastructure to make micromobility rides safe. Cities in the US will need similar investment to reduce car use and encourage smaller modes. "American cities will need to dedicate public roads to micromobility as they have in China, where massive protected bike lanes flank boulevards. Cyclists on 40-pound vehicles should not be forced to ride shoulder to shoulder with cars weighing 4,000 pounds. Asking them to do so is dangerous and impractical. Policy and infrastructure are critical…If American cities follow China's lead and make space for electric scooters and bikes, we can improve air quality and traffic congestion while slashing carbon emissions."[20]

Another low-cost infrastructure fix involves the primary cause of road deaths: speed. Several cities in Europe, including Oslo, Brussels, and Edinburgh, have begun to embrace speed limits for car traffic on city streets of 30 kph or 20 mph maximum speed to promote a more harmonious coexistence between scooters, bikes, and cars. Similar limits of 20 mph in US cities have also been proposed. These speed limits are designed to protect vulnerable road users and allow for the coexistence of modes. In 2023, Transport for London (TfL) reported that a 20 mph speed limit on key London roads led to a 25% reduction in collisions and serious injuries and that collisions involving vulnerable road users decreased by 36% and collisions involving people walking reduced

---

[19] Euwyn Poon and Derrick Ko, "Op-Ed: Now Is the Time for Low-Cost Infrastructure Projects That Can Reorient Our Cities Away from Cars," *Streetsblog USA*, Aug. 14, 2020.
[20] Levi Tillemann, Anthony Eggert, "Is American Micromobility a Bust?" *Wired*, June 10, 2019.

by 63%.[21] Cities in the US should also adopt lower speed limits in downtown areas to improve safety for non-car modes of travel.

By building this infrastructure, cities can encourage more micromobility riders and better achieve their goals to reduce car traffic and encourage multimodal transportation.

We'll Always Have Pari-

One city in Europe exemplifies the arc of regulation and public debate that has challenged the shared micromobility industry from its inception. From 2007, with the launch of its Velib docked bike-share system, Paris has led the way in shared micromobility services. Not only did Paris experiment early with shared fleets of small vehicles, but it also did what few other cities have managed: completely reshape the city's infrastructure to encourage bike and scooter use. Paris has built extensive networks of micromobility lanes and other infrastructure to encourage massive adoption of micromobility throughout the city. Streets once known only for car traffic are now filled with bikes and scooters zipping about freely.

Paris had one of the earliest shared scooter programs, once allowing so many operators and scooters in the city that it wreaked havoc and clutter everywhere, causing the city to revise its program and limit services to three operators. Moving from an open system to a tightly controlled permit program brought more order to the streets. Shared scooter programs were enormously popular in Paris—and that enormous success was also their political downfall. TechCrunch reported that "90% of [Lime's] fleet in Paris is used every day. In 2021, over 1.2 million scooter riders, 85% of whom were Parisians, took a total of 10 million rides across Lime, Dott and Tier. That's around 27,000 rides per day,"[22] but all those rides led to the constant challenge to keep the vehicles parked in proper spots and to deter riders from riding on sidewalks and pedestrian areas. Those who rode scooters loved them, but Parisians who did not ride scooters really hated them. High usage by tourists and

---

[21] Intelligent Transport, "TfL Data Shows 20mph Speed Limits Improve Road Safety in London," Feb. 14, 2023.
[22] Bellan, Rebecca, "Paris Votes Overwhelmingly to Ban Shared E-Scooters," *TechCrunch*, Apr. 2, 2023.

young people led to complaints from older residents about clutter on sidewalks and dangers to pedestrians.

Ultimately, Paris became a story about politics, not transportation policy—with elected officials facing backlash from residents who do not ride scooters but who do vote. The demographics of scooter users skew younger and male, and the key voters in cities often are not. While everyone can ride in or benefit from cars, not everyone in cities wants to ride a scooter and the initial havoc of too many scooters in Paris only served to turn popular opinion against the industry. All of the infrastructure Paris has built for micromobility has been trumped by a political debate that became too heated for even the micromobility-friendly mayor to withstand.

With the city about to host the Olympics, the mayor turned against shared scooters, calling for a citywide vote to "poll" whether scooters should be allowed. Widely viewed as a vote designed to provide cover for removing the scooter program, the city vote was held only in person and on a Sunday, reducing turnout to just 7.5% of citizens (103,084 voters) leading to a high turnout of older voters who dislike the scooter program. As TechCrunch reported, "Dott, Lime and Tier said in a joint statement that the low voter turnout affected the results of the referendum. Only 103,084 people turned out to vote, which is about 7.5% of registered Paris voters. They blamed restrictive rules, a limited number of polling stations (and thus long lines that dissuade young voters) and no electronic voting, saying the combination 'heavily skewed toward older age groups, which has widened the gap between pros and cons.' Additionally, the companies said the referendum was held the same day as the Paris marathon, and that only residents of the city were allowed to vote, leaving out those who live just outside the city but commute in."[23]

After this vote, the most micromobility-forward city has now banned shared scooter programs from September 2023, creating a setback for the industry and residents who had relied on such vehicles for transportation. The shared operators have added electric bikes to continue to serve Paris residents with

---

[23] *Ibid.*

micromobility options, but the industry and riders are hoping the scooters will return. The city's investment in infrastructure for micromobility has made its bike services more popular than ever, showing that investments in non-car infrastructure can pave the way for change. Most in the micromobility industry believe that Paris will eventually return shared scooters to the streets, perhaps with new rules and regulations, given the popularity of the mode, but for now the lesson learned in Paris is that political backlash can overwhelm all the best regulatory and policy intentions.

Luckily, Paris' about-face on scooters stands in stark contrast to other major European cities such as Rome, Madrid, London, Vienna, and Lisbon—and also is at odds with the position of the French central government and other major cities in France. As the industry grapples with ways to improve infrastructure for micromobility and lessen conflicts between scooter riders and other modes, cities must decide if they will pander to vocal residents who dislike the mode or keep trying to get the regulatory framework right.

Marc Bruxelle/Shutterstock

5

Reset the Scooter Rules

In addition to building infrastructure such as protected lanes and parking, cities can also promote micromobility by changing regulations to meet city goals in ways that keep the program financially viable for operators. Micromobility is an important mode for cities, and cities should revise their regulations to allow micromobility to succeed.

Scooter Economics

Some urbanists have questioned why cities should care whether micromobility operators stay in business at all—and criticize the fact that the public right of way is used by private companies. This approach is backwards, as shared operators are now providing for no charge to cities a service that cities previously paid operators to provide for residents. With docked bikeshare systems and Social Bicycle's Smart Bikes after that, cities bought fleets of bikes and/or docked systems from operators for tens of thousands of dollars just to be able to offer residents a new mode of transportation. With dockless fleets, cities get the benefit of a great mode of transportation for residents without paying anything at all. The capital expense of purchasing fleets of bikes and scooters is now borne entirely by private industry. The only cost to cities is providing the light infrastructure of parking spots and bike lanes in the public right of way to allow services to operate. These are amenities that also serve privately owned bikes and scooters and help car traffic move smoothly. Yet, cities turned the tables and started charging fees for operators to obtain permits and imposing new rules that come with hefty price tags for companies.

How do city fees and rules affect scooter economics? After years of experiments in cities around the world, scooter operators have learned the key levers that drive sustainable operations and have issued some recommendations as "best practices for regulating micromobility programs."[1] These recommendations from the industry are designed to help cities create sustainable programs that balance the needs of cities and operators.

A successful shared program must allow each operator to generate sufficient revenue to offset the fixed costs and variable labor costs of the market operations. As discussed below, cities can ensure sufficient revenue in a market by limiting the number of operators and allowing a large enough fleet size based on expected utilization to make the number of trips and revenue per trip attractive for operators. In addition, cities can open all neighborhoods of the city to riders with contiguous operations and sufficient bike lanes to promote safe riding. The following sections discuss some of the key city regulations that affect revenue and costs for operators.

Number of Operators

How many operators should be granted permission to operate in a city? Should the field be wide open to all who can meet the relevant standards, or limited to just two or three providers? This debate has played out over the last few years as cities have tried different approaches. While open markets were initially attractive for cities to get experience with many different operators and determine which companies were a good fit and for operators to learn which markets could be profitable, the difficulties have largely outweighed the benefits and most cities globally are now moving

---

[1] "Joint Statement of Bird, Lime, Spin and Superpedestrian," in US market May 2023 ("In just a short time, shared electric bike and scooter usage has taken off, providing the strongest challenge yet to personal car use in cities. To ensure these options remain a valuable part of city transportation networks, we combined our expertise to develop recommendations to cities that we view as best-practices for regulating micromobility programs.").

toward a limited number of operators under a permitting or RFP system.

In the early days of micromobility, limiting permits to just two or three operators also led to unintended consequences: Competition was so fierce for the few permits available that cities were able to demand compliance with onerous rules, operating standards, and features, and operators promised to deliver. Many of the promises made could never be kept and could not be sustained financially. This race to the bottom caused operators to incur massive losses and frustrate city officials. As the market has matured and consolidated, operators can no longer promise cities a free lunch in order to win a permit, leading to more realistic permit applications and programs as cities realize an unsustainable set of rules will cause the program to fail. As cities have gained more experience managing programs, and operators have learned the costs involved, scooter programs have developed to balance the needs of cities with the financial realities of the operators.

The recommendations from the major shared scooter operators suggest that limiting the number of operators is key to success in a market. Operators are now less willing to enter cities with a free-for-all approach. Cities can encourage healthy competition and consumer choice, while avoiding too many scooters, by permitting two operators in cities with 1000–2000 scooters total and three operators for more than 2000 scooters. This number of operators allows each operator to have a sufficient number of scooters in the market and still generate enough trips per vehicle per day (TVD) to support its operations.[2] It also prevents over-saturation of scooters in a market and reduces the complexity of the program for cities.

The National Association of City Transportation Officials (NACTO) guide to shared micromobility permitting notes some of the benefits of fewer operators: "With fewer operators, each operator has the potential to capture a higher portion of the market resulting in more trips and higher revenue. With more revenue,

---

[2] "Micromobility Industry Best Practice," Statement by Dott, Lime, Superpedestrian, Tier, Voi and Bird in Europe, December 2022; "Joint Statement of Bird, Lime, Spin and Superpedestrian," in US market May 2023.

operators may have more resources to partner on city goals. Longer agreements give operators a higher level of confidence in the local market supporting more investment in local operations and infrastructure."[3] Cities also find limiting the number of operators more manageable leading to a more productive partnership with the operators in the market. NACTO notes that "the selective permit model, and to a greater extent the single-operator contracts, increase city involvement, control and accountability for outcomes."[4] The city of Baltimore, Maryland, has strengthened its scooter program over the years, moving from four operators to two operators and adjusting the rules to provide more flexibility in deployment zones and fleet size. By working with two operators more closely, the city has been able to achieve more consistent compliance and responsiveness while providing better experience for riders.[5] Likewise, Denver, Colorado, was an early adopter of the limited permit approach, choosing only Lyft and Lime to operate scooters in the city with five-year commitments.[6] Many of the major cities in Europe have also chosen to reduce the number of operators in a competitive RFP process, including London, Rome, Madrid, and Vienna.

In some smaller markets, it makes sense for the city to partner with one operator, rather than dividing the scooters among multiple operators. This allows the city to create a close partnership with the operator to foster city goals, increase ridership, and work to respond to concerns of programs as they arise. Single-operator contracts can also allow cities to experiment with other models to support micromobility and promote non-car modes.

One innovative city example is Eugene, Oregon, a college town near Portland. The city partnered with Superpedestrian to offer scooters on an exclusive basis, with one major caveat: Superpedestrian was required to use the same third-party logistics provider that serviced the city's bikeshare program. While many

---

[3] NACTO, "Shared Micromobility Permitting, Process and Participation," Urban Bikeway Design Guide, December 2022, p. 13.
[4] Ibid.
[5] Baltimore Dockless Program, https://transportation.baltimorecity.gov/bike-baltimore/dockless-vehicles.
[6] Denver E-Bike and Scooter Program, Denver's Scooter and Bike Share Program—City and County of Denver (denvergov.org).

cities insist that operators retain their own W-2 employees, rather than outsourcing labor, this innovative model is one that can work in a single-operator town. The city bikeshare operator benefits from the contract with Superpedestrian by adding to its bottom line, and Superpedestrian can provide scooter service in a more cost-effective way given the relatively modest size of the fleet. The city is not a party to the contract, which was negotiated between the bikeshare operator and Superpedestrian, but the city benefits from two sustainable micromobility programs supporting each other. Are there other places where this type of partnership might work or where the city can contribute other support to scooter programs without direct subsidy? For micromobility to flourish and serve more trips than just traditional bikeshare, cities must work with scooter operators to find ways to make scooters a sustainable business.

Fleet Size and Parity

In addition to limiting the number of operators, cities should also allow each operator to have sufficient fleet size to make the program financially sustainable and useful for riders. How many scooters will be sufficient to serve the population while avoiding over-saturation and clutter? If riders cannot find a scooter when they need one, they will not be able to rely on scooters as part of their daily routine, but too many scooters will cause clutter on the sidewalk and result in unused scooters sitting idle for weeks.

Cities have been extremely cautious with fleet size limits at the expense of rider convenience, making it often the case that riders cannot easily find a scooter to rent when they need one. Operators have asked for dynamic caps that adjust to reflect user demand. If cities want to encourage mode shift and prioritize micromobility, they should increase these dynamic caps once the TVD for an operator exceeds one trip per vehicle per day—allowing a greater number of scooters in areas with high demand. Scooter operators now have six years of experience in various cities and have determined that the fleet size should be determined by area of service and scale gradually with demand and compliance proof points—beginning with an initial fleet of 80–120 vehicles/km^2 and growing to at least 200 vehicles/km², or one scooter per 500 people in the city population.[7] These types of metrics can provide guidance to ensure that fleet size can provide a reliable form of transportation to residents without leaving too many scooters unridden.

The National Association of City Transportation Officials regularly issues guidelines for regulating micromobility in cities. NACTO notes in its "best practices" for cities that "regulating fleet size both supports a robust availability of vehicles, while also ensuring cities have the appropriate capacity and resources to oversee shared micromobility systems."[8] Finding this balance in cities, and making the numbers work so that operators can stay in business and riders can rely on the mode, seems to be a major challenge.

The size of the permitted fleet for each operator compared with the other operators in the city, known as fleet size parity, is also a key factor affecting whether operations are financially viable in a city. If one operator is allowed to flood the city with thousands of scooters, even if not often used, then the other operators will not be able to gain market share and compete in a profitable way. By limiting each operator to the same number of scooters in a city, cities can ensure that all operators have a fair chance to provide service in a way that is sustainable. With each operator offering the same number of scooters, services can compete on other metrics such as vehicle quality, ride service, and other features, not just seeing which operator can flood the market first.

One example where fleet parity has made a big difference is Austin, Texas—home of the South by Southwest conference and a hopping downtown with college kids, tourists, and workers eager to ride scooters to work and play. The city of Austin has evolved its scooter program over time, working collaboratively with operators to solve problems and enhance the program for residents and riders alike. Initially, the city had multiple operators with some of the larger players deploying thousands of scooters in the downtown

---

[7] "Micromobility Industry Best Practice," Statement by Dott, Lime, Superpedestrian, Tier, Voi and Bird in Europe, December 2022; "Joint Statement of Bird, Lime, Spin and Superpedestrian," in US market May 2023.

[8] NACTO, "Guidelines for Regulating Shared Micromobility," Sept. 2019, p. 16.

area, with a total of 15,000 scooters deployed. To control the clutter and ensure each scooter had higher utilization, Austin reduced the fleet size for each operator to 2000 scooters with three operators, or a total of 6000 scooters, beginning in June 2023. This parity allows each operator to run a viable business without flooding downtown to try to attract riders.

In contrast, the city of Chicago gave a significant fleet advantage to a single operator—its bikeshare provider Divvy, owned by Lyft. Chicago allows only Lyft's Divvy system to flood downtown with 1000 scooters,[9] while limiting other providers to just 40 scooters downtown. This disparity in vehicle count makes it very difficult for other operators to generate significant revenue in the downtown areas.

Operating Areas

How to manage downtown and popular tourist areas is a key issue for cities. Operators would naturally put most of their fleet in downtown areas where utilization would justify it in order to get as many rides as possible in these dense areas. Cities want to avoid unnecessary clutter with more scooters than are used and also encourage or insist that other areas of the city are also well-served. Scooter companies agree that it is fair to serve all neighborhoods in the city, but unlike public transit services, they are not receiving government subsidies to provide this service. Equity requirements treat scooters like public transit "coverage routes"—which do not have high ridership but incur significant operational costs. These routes are provided not to make money, but because it is fair to provide coverage to everyone in the city. Scooter operators agree with this goal, but struggle to shoulder the financial burden alone. Often, these neighborhoods need micromobility because the city itself has created transit deserts by not providing adequate public transit in the area. Unfortunately, if there is a low ridership of scooters, residents are not benefiting from the cost imposed on operators to provide service in the neighborhood. Are there better

---

[9] Lyft Multimodal Report 2023, p. 15.

ways to provide service to all neighborhoods than specific deployment zones even in areas where there is no demand? Could cities better promote equity by requiring or funding equity discounts on rides wherever taken in the city, which allows the benefit to accrue directly to riders who wish to use scooters, rather than imposing costs on operators for scooters to sit unused in various areas of the city? These are challenges operators can work with cities to address if the program rules allow for sufficient revenue and lower costs overall.

The city of Chicago has implemented a system designed to avoid over-saturation in the popular downtown Central Business District and to spread scooters equitably across a larger swath of the city—all laudable goals, but the impact on operators has been financially unsustainable and disproportionate to any benefit received. The total number of scooters each operator can deploy in the Central Business District is tiny (approximately 40 scooters), and the operator must deploy 4% of its fleet in each of the 10 separate equity sub-areas the city created. In total, Chicago requires more than 50% of an operator's fleet to be deployed in these zones, which experience low ridership, far more than any other city (e.g., 20% in Los Angeles and 10% in Seattle). As a result, in the first year of Chicago's program, the cost to operators to provide a ride was approximately $13 versus only $4 in Los Angeles and $3 in Seattle. Chicago also requires scooters to be repositioned if not ridden in five days, levying fines for scooters not moved on time. With scooters spread out in so many areas with little demand, the labor cost to move scooters is high and not offset by any revenue for those vehicles, leading to further losses.

London operators face a different operating area challenge—the city has not issued citywide regulations permitting micromobility use but has instead allowed each borough to determine its own rules. London operator Lime says the Ultra Low Emission Zone (Ulez) will not reach its full potential without citywide regulations for shared e-bike and e-scooter services. This comes as the Ulez has been expanded to include all of London's boroughs. "London is at the forefront of implementing policies to mitigate the effect of emissions on the climate, and the Mayor's strong commitment to

Ulez is a positive step in drastically minimising the environmental impact of cars in the Capital," Lime's General Manager Manish Kharel told Zag Daily. "But to fully reap the benefits of Ulez and reduce car reliance, a standardised framework for e-bike parking and riding throughout London is of vital importance." Currently, e-bike and e-scooter parking is dictated on a borough-by-borough basis causing confusion among riders on which rules apply where. "Moving to one single framework with 10,000 new parking bays will simplify the process of taking trips across boroughs, improving the likelihood of a significant shift towards alternative modes of transport as Ulez expands. Riders need to be able to end their journey in a tidy manner, but in the location they wish to travel to."[10] London is one example of the negative effects of not providing a sufficient operating area to allow for micromobility use as a reliable form of transport throughout the city.

Permit Length

While many cities started with one-year permits, cities and operators now realize that longer-term commitments on both sides can be beneficial for cities and more economically viable for operators. The shared operators now recommend a minimum length of two years for initial pilots and three to four years for permanent programs.[11] This length of contract allows the chosen operators time to set up and evaluate the program in a city and also provides long-term confidence for the public to adopt micromobility and for operators to make longer-term investments in facilities and people. Permit wars are expensive and time-consuming, and limiting the permitting process to once every three or four years is much more sustainable.

Longer-term permits can also help operators and cities work together to improve micromobility infrastructure. In the future,

---

[10] Alyssa Harris, "London-Wide Shared Transport Regulation a Must to Reap Ulez Benefits Says Lime," *Zag Daily*, Aug. 30, 2023.

[11] "Micromobility Industry Best Practice," Statement by Dott, Lime, Superpedestrian, Tier, Voi and Bird in Europe, December 2022; "Joint Statement of Bird, Lime, Spin and Superpedestrian," in US market May 2023.

cities could also encourage operators with long-term contracts to work together to set up infrastructure throughout the city for parking, charging, and other operational needs to reduce operational costs and trips needed to service micromobility operations.

Selection Criteria

Permit selection should be based on demonstrated excellence in operations and compliance—not on pay-to-play terms where operators can win a permit by offering to share their revenue with cities, or offering low fees to riders that create unsustainable market conditions. Recently, selection criteria in the West of England and Essex served as a deterrent to some operators who declined to participate in unsustainable revenue-sharing arrangements. For example, Lime reportedly withdrew from the Essex e-scooter tender in the UK because 30% of the weight of the decision was based on a revenue share requirement. Hal Stevenson, Director of Policy UK and Ireland, told Zag Daily: "What we've seen in the UK in the last six to 12 months is an increasing trend in revenue share requirements of tenders by which local authorities require operators to split the revenue they generate from the service. It results in a financially negative cycle for operators."[12] Instead, operators should be judged by experience in providing reliable and safe quality of service in compliance with city rules.

Permit selection criteria should also not be dependent on specific technologies or features. Cities should specify desired outcomes in the requirements and remain neutral on specific technologies, features, and operational practices used by different operators.[13] This allows operators to bring experience and creativity to provide great service and curb negative externalities using innovations that improve over time. Whether the technology is swappable versus larger fixed batteries, parking compliance

---

[12] Alyssa Harris, "Revenue Share Requirements Creating a 'Negative Cycle' for UK Operators," *Zag Daily*, Sept. 6, 2023.
[13] "Micromobility Industry Best Practice," Statement by Dott, Lime, Superpedestrian, Tier, Voi and Bird in Europe, Dec. 2022; "Joint Statement of Bird, Lime, Spin and Superpedestrian," in US market May 2023.

technologies, or efforts to enforce geofences, companies should be allowed to continue to experiment and improve outcomes while balancing costs. If cities fix the technology or hardware in the permit and regulations, then companies will not be able to innovate and find new solutions.

Regulatory Burden

As cities have added more and more requirements and regulations to scooter programs, the cost of complying has outweighed the revenue in many markets, leading operators to stop providing service. Even in markets where operators remain, the price of riding a scooter has skyrocketed as companies try to make the unit economics work while meeting the regulatory requirements imposed by cities.

When the story of the micromobility industry is told, there will be blame on the industry for spending its capital too quickly and burning through cash to reach market share, but equal blame will fall on overzealous city regulations that caused much of the burn in the first place. From unprecedented insurance and indemnification requirements, to high city fees, to labor costs needed to respond immediately to misparked vehicles and distribute vehicles in far-flung neighborhoods, cities have imposed burdens on the smallest vehicles in town that they would never dream of imposing on cars. By handing out a small number of permits like medallions, cities forced operators to comply with greater and more costly rules every year to win favor, while missing the chance to take mode share away from cars by promoting and supporting micromobility options.

The industry has faced numerous setbacks and consolidation as the cost to operate in cities keeps going up while revenue opportunities have been curtailed. Now is the time for cities to reverse this trend and partner with shared operators to save this important mode by reducing costs and increasing rides. The following sections discuss some of the key costs that cities can reduce.

Permit Fees

Cities should reduce the permit fees they charge scooter companies. Currently, fees are extremely high for such small vehicles with great road geometry. Why are cities taxing scooter services with high fees, when they serve city goals by providing some of the most efficient and clean transportation available? With shared micromobility, cities get the benefit of an alternative mode without paying the costs of docked infrastructure or purchasing fleets. Micromobility companies currently are paying to provide the vehicles and the service, all at no cost to the city. In light of this benefit, cities should not require shared scooter operators to pay any fees to the city at all.

If any fees are charged, they should be minimal. Any fees charged should be limited to fixed fees paid monthly commensurate with the low fees charged to bikeshare, or per ride fees that scale with usage and revenue.[14] Charging higher fees, or requiring fees to be paid for the year in advance, forces operators to abandon markets as the cash burn is not sustainable. Today, city fees are imposed per scooter, per operator permit, per ride, and for violations of various rules resulting in the impounding of vehicles or parking fines. Cities have justified these fees by claiming they are needed to cover the cost of running the scooter permit program and overseeing compliance with the rules, but it seems unlikely that the costs are anywhere near the millions of dollars of fees being charged.

For example, Chicago collected two years' worth of fees upfront on the first day of the program, before a single dollar of revenue on those scooters was earned.[15] License fees for operators include a $1 fee per vehicle deployed per day. With a standard fleet of 1,000 e-scooters, this equates to $365,000 in additional overhead per year for every operator. Publicly available data show that ridership is around 25,000 trips per month for each operator, with standard pricing and an average per trip cost of $6. Even with a highly

---

[14] *Ibid.*
[15] City of Chicago Rules, Chicago Scooter License Rules and Regulations - Revised Draft (06-23).pdf, p. 21.

optimistic revenue estimate of $1.8M ($150K month × 12), such program fees equal over 20% of total revenue before considering taxes, labor, and operating costs.

Current fees are disproportionately high relative to the road geometry of micro-vehicles and the value to cities in a multimodal system. These fees fail to consider the positive social, economic, and environmental benefits that micromobility can provide. In contrast to the fees charged to micromobility operators, car drivers do not pay anywhere near these regulatory fees, including registration and gas tax, and neither do ride services. Scooter fees are significantly higher. For example, Bird has noted that in the US, "ride-hailing giants like Uber and Lyft pay an average of $0.08/mile to cities, while micromobility operators are charged anywhere from $0.06-$0.45/mile."[16] If cities are concerned about congestion and climate change, they should impose more onerous fees (road pricing) and requirements (limit parking and allocate more space to other modes) on cars and eliminate or reduce fees for shared micromobility devices.

The city of Denver has taken an interesting approach to its program, designed to meet city goals but also be sustainable for operators. After allowing multiple operators in its early program, Denver moved to a two-operator model in 2021—signing contracts with Lyft and Lime with numerous requirements in exchange for five-year contracts and only two operators. In exchange, Lyft and Lime agreed to partner with the city to install and maintain vehicle docking stations and infrastructure in the right of way to support their micromobility services.[17] Denver does not require Lyft and Lime to pay fees to the city and allows them to solicit advertising sponsorship fees. The two operators agreed to provide service to equity zones with 30% of the fleet deployed in specified areas and to invest in parking education and infrastructure to support proper usage of both bikes and scooters. Combined with the city's extensive commitment to implement over 125 miles of bike lanes over

---

[16] Bird, "The Next Normal in Urban Mobility Doesn't Look Anything Like the Old Normal," Bird Blog, June 23, 2020.
[17] License Agreements, Denver Shared Micromobility Services, with Lyft Bikes and Scooters LLC and Neutron Holdings (Lime), May 2021.

five years, the parking infrastructure added by the operators on the public right of way has encouraged proper micromobility use in Denver. By giving scooters a place to ride and a place to park, Denver is serving its goals to encourage multimodal transportation and reduce reliance on cars. The economic terms of the program allow the operators to provide high-quality service across the city without the massive losses required by other city programs.

Labor Costs

In addition to reducing fees, cities should understand and reduce the impact of regulations on the labor costs per trip for each operator. The variable costs of labor are significant and in the US are generally about 50% of the cost of each trip served. Time is money in scooter operations, quite directly it turns out. Operators measure the labor minutes per trip in a market by dividing the total number of scooter trips in a market that day by the total number of labor minutes worked to understand the cost of providing a trip. Each operator must be able to keep the labor minutes per trip to a minimum in order to reduce the variable cost of providing trips as weighed against the revenue earned in the market. The cost of labor associated with deploying, retrieving, charging, and repairing vehicles is significant, and if additional time is spent rebalancing, moving, and re-parking scooters to meet regulations rather than customer demand, the labor minutes involved can quickly eat away any margins.

For example, city rules that require workers to pick up and move scooters for reasons other than customer demand impose costs without a corresponding uptick in revenue. Examples of such rules include moving scooters after 48 hours, rebalancing the fleet on a particular schedule, or distributing a few scooters to many different designated sub-areas of the city. In a dense downtown area with high demand for trips, scooters need rebalancing less frequently and are generating more trips per vehicle per day (TVD). In contrast, rules that require deployment in areas where data show no demand for scooter rides result in extra labor minutes, particularly if the scooters must be moved after a set period of time when not used. If two van trips are required, first to deploy and then to move/re-balance scooters, with still no rides taken, this cost of labor is not offset by revenue. Is the extra cost of requiring operators to move and re-deploy scooters worth it? Ultimately, scooter companies have to pass these costs on to riders if they are to maintain operations in cities with these onerous rules—and the higher cost of rides makes car options more attractive in comparison. If cities want to incentivize micromobility over cars, the rules that increase costs must be weighed against the overall harm to the program and affordability of the mode to residents.

Equity Zones

Where deployments are needed in less dense neighborhoods to equitably serve all neighborhoods, who should pay the cost of making the scooter service available when the customer demand is not sufficient to cover the costs? How can cities best partner with operators to find a way to economically serve the entire city? Fleets should be allowed sufficient numbers of scooters in dense downtown areas where rides are more frequent to offset the service costs in areas with fewer rides. Labor minutes per trip in busy downtown areas can provide more revenue to help subsidize scooters in less frequented neighborhoods. Cities should also remove all fees for scooters to lessen the financial burden of providing citywide coverage in less dense neighborhoods.

While cities and operators both want to serve equity zones well, unlike public transit where "coverage lines" are subsidized by the government, operators receive no such subsidy to serve areas that are less dense or have fewer riders. Cities can support these trips by eliminating city fees and by reducing the regulatory burden of complex zones and rules. Rather than forcing scooters to be left on streets with low demand, cities could allow operators to provide discount ride incentive programs to low-income residents instead, using scooters throughout the city. All neighborhoods should have access to micromobility, but cities should waive fees and requirements to make such service more sustainable and balanced by fleet size in higher-demand neighborhoods.

Parking Compliance and Fines

Cities have imposed several costs on operators to punish operators for the bad behavior of riders in parking scooters. These costs consist of increased labor costs to move or retrieve scooters not parked correctly, often within very tight time windows, and actual regulatory fines for misparked scooters. Both the labor costs and the fines are significantly more than the value of the rides taken on scooters, making the costs disproportionate to the mode. Operators have required riders to submit end-of-ride photographs to encourage them to park in the correct location, including in marked parking zones, locked to street signs or racks, or with sufficient clearance not to block a sidewalk, but scooters can tip over or be vandalized, and not all riders properly end the ride. The question of how to encourage better compliance by riders has been answered by inflicting massive fines on shared scooter operators in cities all over the world.

Cities and operators need to work together to improve parking compliance rather than using punitive back-end enforcement costs that increase the costs of the program without achieving significant upfront improvement by riders. Operators and cities should continue to work together to create sufficient parking infrastructure and help riders understand the rules and how to park scooters and bikes responsibly. As Haya Douidri of Superpedestrian notes, "In cities, the issue of bad parking is twofold, stemming from both the lack of knowledge about parking rules and the insufficient parking infrastructure, especially in densely populated areas. To tackle this challenge comprehensively, we must empower all users with a comprehensive understanding of local parking regulations through campaigns, training events, and digital tools. Simultaneously, investing in adequate parking infrastructure and optimizing parking bay density becomes essential to foster responsible parking practices and create a seamless urban mobility experience for all." While technology can help to some degree, as can human parking patrollers, these are expensive solutions that may outweigh the cost of the ride and must be combined with rider education and sufficient infrastructure to have the desired impact. Additional investment upfront by operators and cities can encourage responsible parking in more efficient ways with education and infrastructure.

Rather than focusing on expensive enforcement solutions, cities and operators should work together on signage and education campaigns to encourage riders to park responsibly in the first instance. Issuing expensive tickets and fines to operators imposes further costs, which even if they can be passed on and collected from riders are another tax on the industry that discourages riders. Parking fines, including impoundment of scooters and time-based fines if scooters are not moved within an hour or two of a report through the city's 311 line or other notification, are unsustainable for operators. These fines based on the actions of riders not operators can quickly evaporate margins. Cities often require or suggest that these costs be passed on to riders to deter bad parking behavior. However, rider behaviors can be difficult to change and higher costs will deter mode shift.

Operators have engaged in extensive marketing campaigns and in-app education to inform riders of legal spots to park the scooters, and cities have also created parking zones in street spots or sidewalk areas painted to indicate proper scooter parking. Some cities in Europe such as Lisbon and Vienna have imposed parking fines as high as 25 euros for each scooter improperly parked, sending piles of tickets to operators to pay. In addition to the regulatory fines, which may be difficult to collect from a rider who only paid five euros for the scooter ride, parking enforcement rules often require additional staff and vans to be deployed quickly to move scooters once they are reported. These additional labor costs add up and detract from regular deployments and maintenance tasks.

When Bird chose to leave San Francisco citing the high costs of operations, the SF Chronicle noted the steep fines on scooter operators compared with fines for similar violations for cars: "Leaving a scooter on its side or not properly locked to a bike rack can cost the company that owns it $150, a fine that can double if it's not moved in two hours. Even in the middle of the night. Parking a car across a sidewalk, blocking passersby far more than a scooter, risks just a one-time $108 fine. In fact, parking a car in a fire lane, a crosswalk or an intersection won't cost as much as badly parking a scooter. And scooting on a San Francisco sidewalk instead of in a bike lane or road can earn a fine of $500, even if there's no bike lane available. That fine is split between the scooter company and the scooter rider. Driving a car in a prohibited space in some cases — even through a parade! — will cost you $100, or a fifth the cost of scooting on a sidewalk."[18] Bird's vice president of city growth and strategy Maggie Hoffman noted that "San Francisco's fines are five to six times higher than in any other city in which Bird operates. San Francisco has the most onerous regulations and is the most difficult to operate in of the hundreds of markets we operate in globally."[19]

While the early days of micromobility were characterized by operators losing money as they offered more rides in more cities even where the program was unsustainable, today's operators cannot afford to lose money in any markets. Reducing the regulatory burden on operators to lower costs is key to promoting micromobility as a key part of multimodal travel in cities.

Rider Restrictions

Cities should also remove other restrictions on scooter riding that drive up the cost of rides without a justifiable benefit. People should be able to use shared micromobility "with minimal restrictions on where and when they case use it" as is the case for privately owned bikes and scooters.[20] One test for whether operational constraints on shared scooters make sense is to look at rules governing bike ridership. As Lime has advocated: "Scooters should be able to ride anywhere and anytime that a bike can ride; subjecting scooters to

---

[18] Heather Knight, "Another Company Leaves S.F., Blaming 'The Most Onerous Regulations' in the World," *SF Chronicle*, Feb. 17, 2023.
[19] Ibid.
[20] "Micromobility Industry Best Practice," Statement by Dott, Lime, Superpedestrian, Tier, Voi and Bird in Europe, December 2022; "Joint Statement of Bird, Lime, Spin and Superpedestrian," in US market May 2023.

stricter rules amounts to regulatory discrimination and can cause confusion among riders, non-riders, and law enforcement. Cities should set the same speed limits or no-riding rules-and then enforce its rules on riders, not operators, to change their behavior."[21] If a rule would be viewed as too heavy-handed for bike ridership, why impose it on scooters, which are more accessible to a greater number of riders? What happens when different modes and form factors such as seated scooters or pods begin to appear, where will they fall in the regulatory scheme? It makes sense to create a single set of rules that apply to all micromobility vehicles, regardless of form, so that riders know the rules and they can be easily enforced against riders not companies when an infraction occurs. Here are a few restrictions that limit the utility of shared micromobility.

Speed Limits. The speed limits imposed by cities are sometimes as low as 5–10 mph, making scooters ineffective and often dangerous to ride alongside much faster vehicles. Scooter speeds should not be capped lower than the 15 mph (24 kph) that bicycles typically ride, allowing scooters to ride safely with the speed of other traffic. At lower speeds, riders will seek to ride on the sidewalk.[22] Companies are also required to use technology to limit speed in certain areas, often confusing riders when the scooter slows down unexpectedly and adding complexity and cost to operations in each city. Cities should limit the areas where speed is regulated solely to highly pedestrianized areas, such as Baltimore's busy Inner Harbor Promenade, while maintaining normal speeds in bike lanes next to such areas to encourage the use of the street space available. As NACTO notes, "reduced speed zones are not a substitution for investing in dedicated bikeways along busy sidewalks or shared use paths."[23]

---

[21] Adam Kovacevich, Head of Government Relations (Americas) for Lime, "Op-Ed: How Cities Should Regulate Scooters," *Streetsblog NYC*, Oct. 29, 2019.
[22] Tanya Mohn, "The Good News/Bad News for E-Scooters and Speed," *Forbes*, May 17, 2023 (citing Insurance Institute for Highway Safety study showing restricted speeds made riders 44% more likely to use sidewalk).
[23] NACTO, "Shared Micromobility Permitting, Process and Participation," Urban Bikeway Design Guide, Dec. 2022, p. 13.

Curfews. In many cities, curfews prevent users from renting shared scooters at night, forcing operators to use technology to limit scooter riding, and reducing the value of the vehicle fleet to riders. This ban on riding at night limits ridership revenue and imposes operational costs. It also makes scooters less useful for those who need to get home at night. People who own bikes or private scooters are allowed to ride them at night. Shared scooters should be no different.

Geofencing. Cities also impose geographic restrictions on where shared scooters can ride and park. Cities should set a single rule for all bikes and scooters, whether owned or shared, and enforce it by ticketing riders who violate the rule. It makes no sense to ban shared scooters from places where owned scooters can ride, and as new form factors emerge in the space and blur lines, a single standard that is easy to follow is essential.

Sidewalk Riding

Cities have sought ways to enforce a ban on sidewalk riding using technology, which is possible but requires extensive time to map, implement, and educate riders to successfully deploy. It also can create a hazard if the technology is used too aggressively to physically slow or stop a vehicle during a ride, as opposed to notifying the rider of undesired sidewalk riding. One of the main benefits of tracking sidewalk riding using technology is to enable cities to implement more bike lanes in the parts of the city where riders seek the safety of the sidewalk due to the lack of bike lane infrastructure. Superpedestrian's recent roll out of its pedestrian defense sidewalk technology in Nottingham, UK, had some interesting results. Nearly 85% of rides did not involve any sidewalk—or pavement–riding in Nottingham. With every ride of additional experience a rider had, his likelihood of riding on the sidewalk diminished, with nearly all sidewalk riding happening with new riders. The results also showed that the city can implement additional bike lanes and infrastructure to help riders feel more comfortable riding and reduce the spots in town where they ride on the sidewalk.[24]

Cities seeking to impose technological fixes to sidewalk riding should partner with operators to subsidize hardware upgrades, pay for detailed maps, and/or provide exclusive operator permits in exchange for such an investment. Imposing such requirements without the necessary time and investment for successful implementation is both dangerous and unsustainable. For example, when the City of San Diego insisted scooter operators implement sidewalk riding restrictions immediately in 2022, all of the operators other than Bird chose to stop operations. "San Diego's year-old crackdown on electric scooters has had the unintended consequence of shrinking usage by 80% and prompting three of the four companies with city scooter contracts to cease local operations. The number of annual rides since the new rules were approved in August 2022 plummeted from 3 million to 595,000 and Spin, Lime and Link have all left town, leaving Bird as San Diego's lone scooter company… City officials concede the sharp reduction in scooters is much larger than anticipated under the new rules, which mandate a version of sidewalk speed throttling that isn't required anywhere else in the world."[25] San Diego chose draconian technological measures without the appropriate time and resources to make the technology work, showing how onerous rules can make scooter programs unsustainable for operators.

Cities need attractive alternatives such as shared micromobility systems to supplement public transit as non-car options. If cities want to continue these programs, they must work with the industry to reform the rules to allow for sustainable and effective operations.

---

[24] Superpedestrian, "Pedestrian Defense Data + Outcomes," Dec. 1, 2022.
[25] David Garrick, "San Diego's scooter crackdown has shrunk usage by 80 percent, raising concerns about climate goals," *San Diego Union Tribune*, Sept. 16, 2023.

Part III
Make Public Transit Great

Buses are as sexy as Amish dresses. American commuters regard them as a poky transportation throwback that they left behind with their last school bus ride. In many large American cities, they're seen as the transportation choice for people who have no choices because they can't afford the expense of a car.[1]

—Janette Sadik-Khan

---

[1] Sadik-Khan, *Streetfight*, p. 234.

6

Fix Buses First

Historically, the only option cities had to improve road geometry was to insist that everyone ride a large city bus. However, city transit buses do not provide the three benefits of cars that people want. We know that people prefer point-to-point travel, on their own schedule, and with the convenience and privacy of traveling alone. Large buses meet none of these needs. Cities can only overcome this problem by offering something better: a faster ride that avoids traffic, on frequent reliable schedules, at a price much cheaper than driving. Public transit must be a great option, not one of last resort. Transit services in the US fall far short of this goal today, especially compared to the superior services offered in Europe.

The Public Transit Struggle

Thus, it is no surprise that we have seen public transit ridership decline for decades in most American cities. As Nicholas Dagen Bloom recently recounted in *The Great American Transit Disaster*:

> Transit riding is a needlessly draining and frustrating experience in most American cities. Even at peak times, buses and trains rarely run when and where everyday riders need to go, including between the suburbs where most Americans now live, work, and shop....To make matters worse, travel by transit usually involves

> time spent in an uncomfortable environment, such as bus stops without shade or rain protection....Substandard bus and train services help sell cars in the United States, with record car ownership partly driven by the lack of a decent alternative.[1]

Transit services need to focus on what they can offer to make buses more attractive, for all riders not just captive riders. Increasing the frequency, speed, and reliability of buses can improve ridership. If a bus sits in traffic just like a car commute, why ride the bus? It must offer something faster than other commute choices. Giving buses their own dedicated lanes that avoid car traffic and have signal priority and improving frequency and reliability are key to incentivizing ridership.

If particular routes cannot offer timely service and are not attractive, cities should experiment with other options. For some routes and times, smaller shuttle buses that operate on demand may be better than a one-size-fits-all approach. Microtransit pilots have faced difficulties in many cities, but that does not mean that the concept cannot work as a substitute for low-ridership routes at off-peak times. More work is needed to continue to experiment with different options. Cities will need to embrace new technologies and consider how to best incorporate new ideas into old services.

Don't Make Other Modes Worse

Some cities have responded to the decline in public transit by trying to make new technology alternatives for commuting worse—in order to deter "competition" with transit—rather than by making public transit itself more attractive. Nearly every new technology has faced the criticism that it pulls riders away from transit, or pulls high-income riders from transit leaving fewer riders and those

---

[1] Nicholas Dagen Bloom, *The Great American Transit Disaster* (The University of Chicago Press, 2023), p. 1.

least able to pay. Cities have imposed fees and other restrictions, or in the case of scooters actually limiting the availability of the vehicles, all in the name of avoiding "competition" with transit.

This is the wrong approach and reflects a disdain for consumers, riders, and the entire transportation experience. The job of transit is to provide mobility, not to get paying riders on buses by making every other mobility option worse. Moreover, this approach ignores the fact that making other options such as Uber/Lyft and micromobility more attractive creates an environment where people do not need to own as many cars and will use all the different modes to get around. As Robin Chase observed, "once you don't own a car, you will travel by all the modes, including public transit, more – because of pricing, because of convenience, because it's the right trip for the right mode. I feel that as much as any of these new modes are adding to the number of people who aren't owning cars, it can only be good for public transit."[2] She also points out that it is not just people using Uber/Lyft who could be riding transit. "We're looking at the people in Uber and Lyft as if they're the only people who aren't taking transit…but what about all the people who drive their own cars? They also are not taking transit."[3] The only way to entice these drivers to take transit, whether they use personal cars or Uber and Lyft today, is to improve transit service and charge fair prices across all modes.

Fast, Frequent, and Free-ish

The freedom to go where you want and when you want, without consulting a schedule, is one of the key features of car travel that keeps Americans commuting in single-occupancy vehicles. Transit that does not replicate this experience will continue to struggle to achieve meaningful ridership and usefulness in cities. People are more likely to start riding the bus if it comes frequently enough that they do not need to consult a schedule. Transit expert Steven Higashide has explained that "The difference between a bus that

---

[2] Interview of Robin Chase, *Smarter Cars Podcast – Ep. 29*, Oct. 22, 2019.
[3] *Ibid.*

runs every half hour and a bus that runs every 15 minutes is the difference between planning your life around a schedule and the freedom to show up and leave when you want."[4]

However, this freedom is not cheap. Transit consultant Jarrett Walker has noted that "Doubling frequency (that is, halving the headway, say, from 30 minutes to 15 minutes) doubles the operating cost. Each increase in the length of the service day is also a corresponding increase in operating cost."[5] This is true as long as we have human drivers on each bus. You could also reduce the operating cost of each trip by increasing the speed of a bus or rail line. "In general, if you were able to cut the travel time of a service in half – that is, double its average speed – your operating cost would drop by up to half. That's because most operating cost is labor, so it varies with time rather than distance."[6]

This investment in more frequent transit has the potential to transform the experience for many more passengers. As Walker explains, "Frequency and span are the essence of freedom for a transit passenger. High-frequency, long-span service is there whenever you want to use it, even for spontaneous trips. If we want people to choose more transit-dependent lifestyles by owning fewer cars, they will need transit that's there most of the time, and where they'll never have to wait long."[7]

Nor is it enough to promise frequency if the transit system cannot deliver those buses on time day in and day out. Reliability is key. As Higashide has observed, "Frequent bus service promises freedom, but when that bus service is unreliable, it's a false promise…Freeing buses from the chaos of city traffic can do a lot to remove chaos from bus riders' lives."[8] Delays on bus routes tend to snowball, resulting in bunching of buses and large time gaps in service before and after the bunched group of buses. Avoiding delays can improve reliability and speed of service.[9] Recently, in light of pandemic reductions in service, the SF Muni bus operators

---

[4] Steven Higashide, *Better Buses, Better Cities* (Island Press, 2019), p. 23.
[5] Jarrett Walker, *Human Transit* (Island Press, 2012), p. 86.
[6] Walker, *Human Transit*, p. 33.
[7] Walker, *Human Transit*, p. 85.
[8] Higashide, *Better Buses, Better Cities*, p. 39.
[9] Higashide, *Better Buses, Better Cities*, p. 40.

were tasked to simply avoid bunching rather than run on a timed schedule given the need to avoid crowding and further delays with a reduced schedule of buses.

To avoid delays, there are several steps cities can take to improve speed and reliability. Jarrett Walker explains the problem of delays in three buckets: "traffic delay is caused by interference of other vehicles; signal delay is caused by required stops at signals; and passenger-stop delay is caused by stops for passenger boarding and alighting."[10] Each type of delay can be reduced, as discussed below.

Bus Lanes

Traffic congestion delays can be addressed with bus-only lanes in the center lanes that are not impeded by cars at the curb. Buses can run faster and on time when they have their own bus lanes. "The fundamental problem with bus service isn't that the buses are incapable of higher speeds. Even though a single bus can take the equivalent of fifty cars off the road, buses are forced operate at the same creeping pace of crosstown traffic as people who drive private cars."[11] UCLA's Institute of Transportation Studies has called for the use of "tactical transit lanes," defined as "a bus only lane tactically implemented in dense, congested areas to speed up transit without major capital improvements."[12] These lanes can help move buses more quickly in congested stretches on commute corridors, increasing both speed and reliability of service. Both improvements can lead to increased ridership as riders realize that the travel is quicker and more reliable on time.

Bus lanes can either be permanent and enforced all day in all directions, or can be limited to commute hours in one direction only. Bus lanes require less investment than other types of infrastructure. Tactical bus lanes may only require red paint, signage, and striping, or even cones to start a temporary pilot. These options are not very expensive. Bus lane projects are "quick, low-cost and

---

[10] Walker, *Human Transit*, p. 99.
[11] Sadik-Khan, *Streetfight*, p. 237.
[12] UCLA Institute of Transportation Studies, *Best Practices in Implementing Tactical Transit Lanes*, Feb. 2019, p. 1.

reversible" and can often be implemented first as pilot projects to assess the efficacy.[13] In San Francisco, an SFMTA study showed that the addition of red-painted bus lanes in 2014 decreased bus delays and resulted in "a 25% improvement in transit reliability."[14] The SFMTA found that "on its red-painted Third Street corridor, the number of drivers violating transit lanes fell 48 to 55% (depending on the time of day), even as traffic increased."[15]

In addition to new road treatments such as red bus lanes, some cities may want to implement additional route and operational changes to make bus rapid transit routes or BRT. These routes are typically longer and can serve up increased transit speeds beyond the dense urban core and into less dense or suburban areas. These larger projects often include other roadway improvements such as signal priority, streetscape changes, and enhanced pedestrian or bicycle access. Bus rapid transit routes may cost as much as $1 million per mile, compared with tactical transit lanes, which typically run $100,000 per mile, according to UCLA.[16] Starting with a tactical transit lane, even as a temporary pilot project, can often show residents the value of these improvements. San Francisco recently implemented a Bus Rapid Transit project on Van Ness Avenue, designed to prioritize frequency and reliability for customers. The planned improvements are expected to cut travel times for many lines by 32%. Some features of Bus Rapid Transit on Van Ness include "Dedicated transit lanes that are physically separated from the other traffic lanes, for use by Muni and Golden Gate Transit buses only; enhanced traffic signals optimized for north-south travel with Transit Signal Priority, which gives buses the green light as they approach an intersection; low-floor vehicles and all-door boarding, that make it quicker and easier for passengers to load and unload at each stop; safety enhancements for people walking, like sidewalk extensions, median refuges, high visibility crosswalks, and audible countdown signals; and fully

---

[13] Ibid.
[14] UCLA Institute of Transportation Studies, *Best Practices in Implementing Tactical Transit Lanes*, Feb. 2019, p. 26.
[15] Ibid.
[16] UCLA Institute of Transportation Studies, *Best Practices in Implementing Tactical Transit Lanes*, Feb. 2019, p. 3.

furnished boarding platforms that include shelters, seating and NextMuni prediction displays located at key transfer points."[17]

Obviously, bus lanes only work if cars stay out of these lanes, which requires some amount of enforcement. This can include issuance of citations to car drivers blocking bus lanes, as well as driver education and signage. "Painting bus lanes red...has been widely shown to help with 'self-enforcement'; when buses get the red carpet treatment, other vehicles are up to 50% less likely to block the lane."[18] In some states, cities can also issue tickets to cars driving in the bus-only lane using camera enforcement on buses.[19] In the future, more jurisdictions will likely allow for automated enforcement, which should increase compliance.

New York City also banned cars from 14th Street in Manhattan, making it bus-only. "After a successful pilot, the popular 14th Street Busway was made permanent in June 2020, with a plan to extend bus lanes east. The 14th Street Busway pilot...has received international attention, as it has successfully increased bus speeds by as much as 24% and ridership by as much as 30%."[20] Not only did traffic and bus travel times improve, but traffic crashes were reduced. "In the four months since the busway began in October, total crashes are down 53% and injuries are down 63% compared to the same four-month period a year earlier. Crashes that resulted in injuries are down 68%."[21]

While car drivers object that creating bus lanes will cause more car traffic, or increase traffic on other streets, such traffic has not materialized in the places where bus lanes have been implemented, including this recent project in New York. Drivers may think giving buses a lane is bad for cars, but if buses run faster and on time, more people who can ride them will do so. This will reduce car traffic in the remaining lanes. Moreover, bus lanes take buses out of the car lanes where they tend to block traffic each time they stop

---

[17] SFMTA, Van Ness Improvement Project, at https://www.sfmta.com/projects/van-ness-improvement-project.
[18] Higashide, *Better Buses, Better Cities*, p. 42.
[19] Sadik-Khan, *Streetfight*, p. 238.
[20] NYC DOT, "Better Buses," accessed at https://www1.nyc.gov/html/brt/html/routes/14th-street.shtml.
[21] Gersh Kuntzman, "The 'Busway' Proves Another Benefit of Car-Free Streets: Safety," *Streetsblog NYC*, Feb. 17, 2020.

and then merge with traffic. "Once you decide your streets are designed for people movement rather than vehicle movement, turning car lanes into transit lanes not only is fair but is also the most effective way to maximize the total number of people who can move along a street."[22] If cities take away street parking in congested downtown corridors where bus lanes are needed, that space is available without taking away car travel lanes at all.

Signal Priority

Cities can also give buses traffic signal priority at intersections so they can move faster and avoid delay at stoplights. "Using infrared signals, traffic lights can be altered so that buses can proceed more quickly through intersections."[23] Green lights can be extended to let the bus through, or bus lanes can get the green light before other traffic. Traffic signal priority technology requires a little more investment, but is still not very expensive compared with other infrastructure projects.

New York City issued a report on its implementation of "transit signal priority" technology to improve bus speeds. The report stated that "Transit Signal Priority (TSP) is a technology that is capable of enhancing traditional transit services by facilitating bus movements through intersections controlled by traffic signals. On busy bus routes in New York City, buses spend about 21% of their time stopped at traffic lights, which is a major contributor to slow bus service around the City…TSP in New York City involves both passive and active priority systems. 'Passive' TSP means traffic signals are better coordinated, thus improving traffic flow for all vehicles along the bus route. 'Active' TSP requires the transit vehicle to communicate with the traffic signals, in order to dynamically adjust the signal timing in its favor by either extending the green signal or shortening the red signal at the approaching intersection."[24] The report also states that "on average, TSP has reduced bus travel

---

[22] Walker, *Human Transit*, pp. 105–6.
[23] Sadik-Khan, *Streetfight*, p. 238.
[24] NYC DOT, "Green Means Go: Transit Signal Priority in NYC" Jan. 2018, accessed at http://www.nyc.gov/html/brt/downloads/pdf/brt-transit-signal-priority-july2017.pdf.

times about 14% during weekday peak morning and evening commuting periods."[25]

Cities should invest in these technologies to help provide faster buses and more reliable service. It makes sense to give signal priority, as well as lane priority, to buses carrying 50 people with great road geometry. "Signal priority can be justified by a goal of managing a street for person trips rather than vehicle trips. Dumb signals treat all vehicles equally, whereas signal priority gives preference to transit because it represents more people."[26] Compared to the single-occupant car sitting next to the bus at the intersection, the bus is moving 50 times as many people—it deserves to move faster. Every time cities can make improvements to the bus experience, more people will get out of their cars and ride the bus instead, reducing the remaining car traffic.

Boarding/Payment

Cities can also improve boarding and payment systems to make riding the bus easier and faster. More than a quarter of bus delay is due to "dwell"—the time it takes at a bus stop for passengers to board or alight, and pay the fare.[27] Delays in boarding are due to "the limitations of space on the vehicle, and on the sidewalk or platform, as well as the number and width of doors."[28] San Francisco implemented all-door boarding systemwide, including buses, in 2012. The SFMTA report in 2014 found that "legalizing All-Door Boarding has encouraged boarding customers to distribute themselves more evenly between the front and rear doors, thereby reducing average dwell times. Pre- and post-implementation surveys at busy Muni stops found average reductions of 1.5 seconds (38%) per customer entry or exit. All-Door Boarding also has improved dwell time consistency and lowered variability, an

---

[25] *Ibid.*
[26] Walker, *Human Transit*, p. 101.
[27] Sadik-Khan, *Streetfight*, p. 237.
[28] Walker, *Human Transit*, p. 102.

important factor in helping reduce vehicle bunches and gaps and making service more reliable and predictable."[29]

San Francisco also moved to a "proof of payment" system, in which passengers do not have to pay the driver, but are instead subject to random inspection to see if they paid. This avoids passenger delay when people have to queue to board and pay and facilitates all-door boarding. Another approach is to require showing proof of payment when boarding, but to allow payment before boarding. In New York, they set up a system for passengers to pay at the bus stop and get a receipt. With pay stations on the curb, the time for passengers to board the bus decreases, speeding up the whole line.[30]

Another source of passenger stop delay is the number of bus stops on a route. If the bus has to stop every block to pick up and drop off passengers, the delays will be greater than if it only stops occasionally.[31] As stops close together will slow down buses, "optimal stop spacing is as wide as possible while still making it possible for the intended customers to get to the service."[32]

The placement of bus stops can also assist in keeping traffic moving. Cities can improve their street design for both buses and pedestrians by using curb bulb-outs for bus boarding areas. "When buses have to pull over to the curb, they get delayed fighting back into traffic…bus stops can be 'in-lane,' with the sidewalk extended to meet the bus."[33] This small change prioritizes bus speed when the bus does not have its own lane.

Finally, all of these improvements to increase speed and reliability are not much comfort to a potential rider who has no idea if or when the next bus is coming. Transit systems must respect customer time by providing them with up-to-date and accurate information. Customers need to know that they can rely on the information provided and that it is not outdated or wishful thinking.

---

[29] San Francisco Municipal Transportation Agency, All Door Boarding Report, 2014, accessed at https://www.sfmta.com/sites/default/files/agendaitems/2014/12-2-14%20Item%2014%20All%20Door%20Boarding%20Report.pdf.
[30] Sadik-Khan, *Streetfight*, pp. 237–38.
[31] Higashide, *Better Buses, Better Cities*, p. 41.
[32] Walker, *Human Transit*, p. 102.
[33] Higashide, *Better Buses, Better Cities*, p. 41.

Today, technology is available to provide real-time information on bus timing, how full the bus is, and other nearby alternatives. Transponders on buses can ping every 30 seconds to provide up-to-date, real-time information. Knowing your bus is actually going to arrive in five minutes makes all the difference to riders. If you have no idea whether the bus is actually coming, you are more likely to give up and use another mode.

Technology has also improved to allow for LED screens at bus stops and on buses themselves to provide real-time route and bus information. Companies such as TransitScreen have developed signage to give up-to-date information to passengers. Countdown clocks and other signs that indicate when the next bus will arrive can make waiting for the bus less stressful and allow customers to plan their travel if delays are occurring on the system.[34] Transit systems must invest in information systems for customers if they want to reap the benefits of increased speed, frequency, and reliability.

---

[34] Higashide, *Better Buses, Better Cities*, p. 109.

7

Plan for the Future

What will the future of public transit look like? How will transit systems need to evolve over time? One concern is that the development of autonomous vehicles will result in cheap, ubiquitous on-demand ride services that cost little more than a bus ride. How should buses evolve to take advantage of new technologies and provide better service to customers? Clearly, buses will also become autonomous in the future, providing more service for lower cost. Some buses may also run on-demand rather than fixed routes during off-peak hours, or in places where demand is lower. Rather than fighting these new technologies, transit services should partner with companies in this space and explore new possibilities for better service.

Mobility on Demand

For some routes and times, cities cannot offer timely or attractive bus service. Often, this is where the area served is not sufficiently dense, or demand is low at certain hours. Rather than continuing to spend money on unattractive and rarely used services, cities should experiment with other options. For these routes, smaller shuttle buses that operate on-demand may be better than a one-size-fits-all approach. Cities are starting to explore using on-demand models for less-traveled routes and times to find more cost-effective and efficient ways to move passengers.

These microtransit, or flexible transit, services do not serve very many passengers per hour because they do not run on direct, fixed routes. They can play a role in providing "coverage" service in areas with low ridership. Balancing the trade-off between maximizing riders and providing service to every neighborhood is a challenge for cities, and they are looking to technology to help them save money and provide more efficient services.

Transit consultant Jarrett Walker describes the problem as one of contradictory missions: Transit agencies have two goals that are hard to reconcile—(1) "serve all parts of our community" and (2) "maximize ridership with our fixed service budget."[1] The first goal is referred to as "coverage," and the second goal is "ridership." Coverage means the transit agency "must serve everyone in its service area – that is, everyone who pays taxes to the agency and votes for the elected officials who will make decisions about transit."[2] Ridership refers to "deploying service the way private business would, with the aim of the highest possible ridership for a given service budget."[3]

Ridership-focused routes provide service that is direct and serves many passengers per service hour. This type of service helps transit offer a realistic alternative to private car travel. It is also financially viable. In contrast, "coverage" service requires transit agencies to provide service to every part of town, even if very few people actually use the service there. Often, this is due to the lower density of population and the related fact that in less dense areas with more space, people are more likely to own and use cars. Studies have found that regarding the density of housing development most of the time "if you double the density, transit demand goes up by more than double."[4]

Coverage routes may provide some service to geographic areas that have little demand, but the service often is not very good. It is infrequent and the routes are not direct, resulting in long trip times and headways of an hour or more. "This service is inconvenient to use, but it may provide essential lifeline access to people who need service, or be justified by the need to provide service to every municipality that pays taxes into a district."[5] The problem

---

[1] Walker, *Human Transit*, p. 118.
[2] Ibid.
[3] Ibid.
[4] Walker, *Human Transit*, p. 123–24.
[5] Higashide, *Better Buses, Better Cities*, p. 24.

with coverage routes is that they lose money and continue to perpetuate the view that buses provide terrible service and are only for those who lack any other option.

Transit agencies often complain that tech companies such as Uber and Lyft do not have these coverage challenges and concomitant financial burdens. They point out that serving every area of town sometimes limits the amount of funds available to serve most people well. It is clearly time for a new approach to this inherent conflict of goals. Serving everyone should not result in transit serving no one well. This is a good use case for on-demand bus service using smaller vehicles where demand is not high.

Consultant Jarrett Walker has been a vocal critic of these microtransit, or flexible on-demand services, as part of transit agency offerings, but even he agrees that they can be a tool for coverage service. "Only if the goal is coverage do these services ever make sense, so only in that context does flexible service appear as a possible solution."[6] Walker claims that "[f]lexible service will never compete with fixed route on ridership grounds, so it should stop pretending that it can. Market the service as what it is. It's one tool for providing lifeline access to hard-to-serve areas, where availability, not ridership, is the point."[7] Setting aside for a moment the question of whether on-demand services may have any other use cases in the future, it seems clear that replacing fixed-route "coverage" service with on-demand shuttles could provide better service to customers living in neighborhoods where existing bus service has been unattractive and infrequent.

Walker notes that "[t]here is no particular efficiency in the fact that flexible transit vehicles are smaller than most fixed route buses, because operating cost is mostly labor." It seems that this calculation would change with autonomous vehicles, if they can reduce labor cost. So, Walker's criticism that these services will never make sense other than as lifeline service may not hold true forever, as technologies advance.

---

[6] Jarrett Walker, "What Is Microtransit For?" *Human Transit*, Aug. 2019, https://humantransit.org/2019/08/what-is-microtransit-for.html.
[7] *Ibid.*

There may also be other specific use cases today where on-demand shuttle service could be cost-effective for cities. Some cities have partnered to replace city bus service with ride service credits, with varying degrees of success. A report from microtransit provider Via and The Boston Consulting Group's think tank looked at on-demand, microtransit use in Arlington, TX, Berlin, Seattle, and West Sacramento, CA. The report noted that "Our study showed that on-demand transit services work. In the right regulatory context, with lower per-passenger subsidies than those provided to comparable public services, these initiatives can benefit passengers and cities alike. Their convenience and flexibility improve the user experience over fixed-route mass transit while bringing good jobs within reach of neighborhoods poorly served by the status quo. They also generate less congestion and pollution than solo passenger travel."[8] The report noted that "In Arlington, for example, the service helped eliminate nearly 400,000 miles of travel that would have occurred if the passengers had driven solo, corresponding to a reduction of 36% of total vehicle miles traveled."[9]

Los Angeles has also used Via in a partnership to provide on-demand access to transit and recently extended its pilot program. "The pilot program is an experiment in bringing ride-hailing technology to low-income, disadvantaged riders and riders with disabilities who might not otherwise be able to access innovative mobility options to get to and from major transit centers… Since launching, the service has provided more than 70,000 rides and exceeded its key goals in terms of rides per week, rides per driver hour, wait times, and customer satisfaction. More than 1,000 riders have used Via's call-in center, indicating that people without smartphones are using the service. In addition, the service has provided more than 800 rides to passengers requiring special assistance or wheelchair accessibility at about a third of the cost to taxpayers of an Access Services ride."[10]

---

[8] BCG Henderson Institute, *On-Demand Transit Can Unlock Urban Mobility*, at 2; http://image-src.bcg.com/Images/BCG-On-Demand-Transit-Can-Unlock-Urban-Mobility-Nov-2019_tcm9-233418.pdf.
[9] *Ibid.*, p. 4.
[10] Los Angeles Metro, "L.A. Metro Extends Rideshare Pilot Partnership with Via," *Metro Magazine*, Jan. 24, 2020.

Cities will need to continue to experiment to find particular use cases where on-demand microtransit can save money and also provide greater mobility and service to customers. As new technologies are introduced and refined, cities should remain open to finding new ways to combine technology and service concepts to increase public transit usage and reduce costs.

The Future Will Be Automated

Public transit historically has not been able to pay for itself, leading to many challenges around budgets and service. Providing bus and rail service costs more than transit agencies can charge in fares and therefore has generally required government investment or subsidy. Urbanists often lament that public transit is viewed as a "subsidy," while car infrastructures such as highways and bridges are considered "investments."[11] Setting aside the semantics, the financial challenge for transit agencies is real. When cities face budget deficits, or enormous financial and health challenges like the 2020 global pandemic, transit service often must be eliminated or reduced, further hindering the adoption of non-car travel patterns. This tragedy and enormous financial challenge must lead us to consider how technology might reduce the cost and improve the services of transit agencies to alleviate some of the need for government funding.

As private operators such as Uber and Lyft, or micromobility companies such as Bird and Lime, have entered into cities, many urbanists have snickered at their attempts to run profitable businesses, claiming that transportation services can "never make money." While urbanists criticize these "VC-funded tech companies" as a model that prioritizes profit over serving the community equitably, they discount the possibility that technology could save money and ultimately, through innovation, provide better services to all members of the community.

---

[11] Interview of Robin Chase, *Smarter Cars Podcast – Ep. 29*, Oct. 22, 2019 (noting we should use congestion pricing money to improve public transit and the unfortunate perception that we "invest in highways and subsidize public transit.").

Saving money and not having to fight for transportation dollars in public budgets are transformational for both private and public transportation providers. Rather than declaring it impossible to run services that do not lose money, urbanists should work with technologists to find ways that technology can complement urban planning policies to meet city transportation goals.

One of the biggest costs for public transportation is labor—paying humans to drive buses and trains. Many train systems are now starting to use autonomous technology, but we have not yet seen the introduction of autonomous buses, other than trials. In 2013, the average fare for both urban and suburban bus and rail transit in the US was $0.28 per mile. The total operating plus capital cost for all metropolitan transit services was $1.13 per mile ($0.78 operating and $0.35 capital). For buses, the costs were over $1.36 per mile.[12] Thus, bus operation is one area that would benefit the most from the introduction of autonomous vehicles.

As NACTO has recognized, "transit is one policy area where vehicle automation and its precursor technologies can have the most immediate direct impact. Automated technologies are especially suited for predictable, fixed routes. Operators can reduce costs and increase service quantity and quality by shifting or augmenting driving functions with autonomous technology...In the long term, full automation can enable agencies to further expand service."[13] Automation of transit buses and shuttles could lower fares by greatly reducing labor costs. Transit agencies could reinvest driver compensation savings into reduced or eliminated fares, or more extensive and frequent service.[14]

While labor unions and others express concern about reducing operator jobs, this is not a reason to shun autonomous vehicles once they become available. First, the transition will take many years and thus is unlikely to displace a large number of current employees. Second, as with truck drivers, there is currently a shortage of bus operators in many major cities where bus lines are

---

[12] Office of Budget and Policy, 2015 National Transit Summary and Trends, Dept of Transportation, October 2016, cited at *Three Revolutions* p. 212.
[13] NACTO, *Blueprint for Autonomous Urbanism*, p. 46.
[14] Sperling, *Three Revolutions*, p. 141.

cut due to insufficient staffing. For both of these reasons, the perceived threat to bus operator jobs is unlikely to materialize.

Some critics of autonomy have posited that the driver can never be removed from public transit because of the need to have a monitor there to protect passengers, answer questions, facilitate fare collection, and report problems on the line with equipment or schedules,[15] but these functions can and will be replaced with cheaper technology-based options to facilitate the cost savings that autonomous vehicles can provide. In-cabin technology already exists to provide video monitoring and voice interactions to facilitate safe operations. The cost savings available with autonomous technology far outweigh the customer service aspect of having a human bus operator on board—and many of those functions can be implemented by tele-operated camera systems and other two-way monitoring.[16]

Automation might also help improve on-time performance and increase speeds by avoiding delays caused by sick or late drivers and transitions between drivers and driver breaks. Better connectivity among autonomous buses and communication with traffic signals would allow buses and perhaps other vehicles with multiple passengers to travel faster and reduce delays.[17] For all of these reasons, public transit will be a primary use case for autonomous vehicles and fixed routes in major urban areas will implement all different sizes of autonomous buses and shuttles when the technology becomes available.

First-Mile/Last-Mile Solutions

For longer-distance commutes on commuter rail and ferries, many more people would ride public transit if they could get from their homes to the transit stop or station in a convenient, fast, and cost-effective way. Commuter rail generally refers to longer-distance

---

[15] See, *e.g.*, Higashide, *Better Buses, Better Cities*, pp. 111–13.
[16] See Interview of Modar Alaoui, CEO of Eyeris, *Smarter Cars Podcast – Ep. 44*, July 13, 2020 (noting in-vehicle camera systems can provide scene understanding and monitoring of drivers and passengers on public transit and in cars).
[17] Sperling, *Three Revolutions*, p. 141.

services that bring people in from the suburbs to the city center. For example, in the San Francisco Bay Area, it might include Caltrain from the South Bay, BART from the East Bay, and Golden Gate Transit from the North Bay. These serve single-access points into San Francisco over the 101/280 freeways, the Bay Bridge, and the Golden Gate Bridge, respectively. Car traffic through these bottlenecks greatly exceeds the available space during commute hours, and therefore, these commuter rail, ferry, and bus options are popular and crowded.

The only issue for passengers is how to get from their suburban homes to the Caltrain station, the BART station, or a Golden Gate bus stop or ferry dock. This is referred to as the first-mile problem. Public transit stations are often only accessible by car and parking lots fill up quickly, making them an unreliable option. If you drive to a station and there is no place to park, you are stuck. In the suburbs, often the current cost or time to get an Uber or Lyft to the station makes it unattractive as a daily option. If micromobility options were available with protected places to ride, or ride services were cheap and ubiquitous, many people would use them for this purpose.

Some cities have recognized this need for first-mile/last-mile transport and have partnered with ride services to subsidize the cost. For example, some transit agencies are partnering with microtransit providers such as Via to provide on-demand access to transit stops and stations in areas with otherwise infrequent transit service. Sacramento, California, launched what may be the "largest on-demand micro-transit project in the US."[18] The service offers "curb-to-curb transit using 'virtual bus stops' generally within about a block of a rider's origin or destination," and riders are able to transit to the system's other fixed-route buses or light rail for the $2.50 fee.[19]

Los Angeles Metro has also had success partnering with Via. "Ridership has grown from week to week, with only 100 to 200 rides per week in the initial few weeks. As word of the service spread, however, ridership picked up to 3,500 rides a week prior to

---

[18] Skip Descant, "The Latest Transit Trend Is Somewhere between a Bus and Uber," *Gov Tech*, Jan. 31, 2020.
[19] *Ibid.*

the slowdown brought on by the novel coronavirus, which took ridership down to 1,300 rides a week."[20]

Once a commuter does take a train into the city center, sometimes the walk from the station to the office is too long to be reasonable—perhaps 30 minutes or longer. For those trips, the last mile is a problem and a deterrent to using transit. Solving both of these pain points—to and from the transit stop—would make more people willing to use transit options. This is known as solving the "first-mile/last-mile" problem, where riders need help getting from their homes to the train station or bus stop (first mile) and/or help getting from the station or bus stop on the other end to their office (last mile).

Cities would benefit from encouraging micromobility and ride service options to solve the first-mile/last-mile problem for more commuters. Recent studies of existing ride service use in cities have shown that ride service availability has reduced the use of city buses and light rail but has increased use of longer-distance commuter rail services, presumably by solving this first-mile/last-mile problem.[21] Cities should continue to experiment with incentives for mobility services companies to solve these first-mile/last-mile problems.

A Fairer Fare?

As cities design the future of public transit, the question of fare policy will be front and center. If we want more people to ride the bus, why not make it free so that people do not need a fare card or app and can hop on and off as needed?[22] If big cities made the bus free, how would they pay for it and how would they run enough buses during peak commute hours to service the demand? How would they even know how many people are riding the bus if

---

[20] Skip Descant, "Transit Partnerships Take on First/Last-Mile Problem," *Gov Tech*, June 2020.
[21] Clewlow, R.R. and Mishra, G.S., "Disruptive Transportation: The Adoption, Utilization, and Impacts of Ride-Hailing in the United States," Institute of Transportation Studies, University of California, Davis, Research Report UCD-ITS-RR-17-07 (October 2017).
[22] Ellen Barry, "Should Public Transit Be Free? More Cities Say, Why Not?" *New York Times*, Jan. 14, 2020.

you do not have to tag on and off with a fare card or app when you ride?

Los Angeles is one major city looking at this issue to determine whether its metro service should be free to attract more riders post-pandemic. LA Metro's report in May 2023 showed that LA spends $79 million annually to collect $106.5 million in fares, resulting in net revenue of $28 million, which represents a net farebox recovery ratio of just 1.25%.[23] In 2019, it was 14.6%, and by comparison, in 2019 the farebox recovery ratio was 52.6% in New York and 71.7% for BART in San Francisco. Given that most of LA's budget is funded by county sales tax revenues, it makes sense for LA Metro to stop collecting fares. Fare enforcement leads to disparate impacts on people of color, and fareless transit can most help by reducing financial burdens on low-income residents.

Ultimately, cities will need to decide how much to alter fare policies to serve city transportation goals. Some cities may decide to make more radical changes to how they collect fares for different modes. Back in 2020, the head of the SFMTA noted one possible way San Francisco could alter fare policy: "We charge San Franciscans $2.50 for two square feet of space on crowded buses during peak hours. What if instead of charging them $2.50, we paid them $5 for making more efficient use of our mobility system and what if we paid for that by charging those who were taking 300 square feet of roadway space for their convenience?"[24] For instance, passengers riding a bus into downtown might receive a travel credit paying them for riding the bus on that busy road while passengers driving their own cars alone, or riding the same route in an Uber or Lyft, might pay extra fees for that privilege, since cars take up much more road space per person than riding the bus. Implementation of these types of policies in the future could bring equity to fare policy that is based on road geometry rather than the typical assessment of minutes or miles traveled.

Consultant Jarrett Walker has noted that it is difficult to make fares "equitable" in the sense of "fair share of cost" of using a given

---

[23] Joe Linton, "New Report Makes Case for Universal Fareless Transit at Metro," *Streetsblog LA*, May 11, 2023.
[24] Interview of Jeffrey Tumlin, SFMTA, *Smarter Cars Podcast – Ep. 35*, Jan. 2, 2020.

service, and cities should accept fares that will "never be perfectly fair." Instead, he suggests that cities acknowledge "the real purpose of a fare system is to bring in a needed level of revenue while imposing a minimum of delay, hassle, confusion and perverse incentives."[25] By focusing on these outcomes, cities can encourage ridership and maintain revenue to support service.

A report by Transit Center in 2019 noted that "revenue should not be the sole consideration of fare policy" and that "transit agencies should evaluate changes to fare policy in light of other goals."[26] Transit Center cautions transit agencies to avoid fare policies that are confusing or economically regressive, such as transfer fees making connected trips more expensive, distance-based fares that make the total cost of the trip hard to understand, and criminal penalties for fare evasion that are disparately enforced against people of color.[27]

Whether cities can take fare policy a step further and bring concepts such as road geometry to bear will depend on how other modes are priced. To encourage people to use cars less, cities should price road use along with fair pricing for public transit. Even making transit free will not have the desired impact if changes are not made to how we price car rides during peak times. Fare policy is an additional tool that can be used to reflect new priorities.

Cities can start by allocating street space away from car parking and toward bus lanes and protecting space for micromobility. Then, they can use economic tools to price road space effectively and encourage better geometry in congested areas. It will take all of these tools working together to move people away from single-mode travel in cars and toward a multimodal future.

---

[25] Walker, *Human Transit*, p. 145.
[26] Transit Center, "A Fare Framework: How Transit Agencies Can Set Fare Policy Based on Strategic Goals," p. 5.
[27] *Ibid.*, p. 4.

Index

A
American Society of Civil Engineers, 11
Automobiles, "attrition" of, 23
Autonomous technology
　cost savings, 91
　financial and health challenges, 89
　in-cabin technology, 91
　labor cost, 90
　on-time performance and speed, 91
　private operators, 89
　shortage of operators, 90–91
　urban planning policies, 90
Average traffic volumes into cordoned area, 18

B
Bird, 89
Bus lanes, 27
Bus rapid transit routes (BRT), 78–79

C
Car
　bundle of trips, 4–5
　comfort and convenience, 4
　ownership, 5
　regulation, 7
　transportation, 3
　travel, 5
　unbundling of, 5
Car parking, 24
　delivery zones, 25
　"drive and park," 25
　land use, 24
　"ride and drop-off," 24, 25
　ride services and delivery vehicles, 25
　street space allocation, 24

Citywide transportation network, 42
Commuter rail, 91–93
Compliance
　bus lanes, 79
　parking and fines, 64–66
　scooters, 51–54, 58
Congestion pricing, 16–17
　political changes, 20–21
　program, 18–20
Cordon pricing, 16
Curfews, 68

D
Delivery services, 29
Disrupting walking, 35
Docked bikeshare system, 40, 49
Downtown street parking, 27

E
e-bike, 56, 57
e-commerce, 29
Electric bikes and scooters, 28–29
Equity, 55, 56, 61, 63–64, 94
e-scooter, 56, 57
Essex, 58
EVP of Policy and Strategy at Superpedestrian, 36

F
Fare policy, 93–95
First-mile/last-mile transport, 91–93

G
Geofencing, 38, 68
2020 Global pandemic, 89
Global Positioning System (GPS), 13
GPS-enabled dongle, 13

I
Induced demand in transportation, 14
Infrastructure
　congestion, 14
　docked infrastructure/purchasing fleets, 60
　education, 65
　invest in, 33–47
　longer-term permits, 57
　operators and cities, 64
　parking spots and bike lanes, 49
　signal priority, 80

L
Lime, 61, 89
Local charging, benefits of, 41
Lyft, 7, 55, 61, 75, 87, 89, 92

M
Market operations, 50
Micromobility, 23
　autonomous technology, 89–91
　bike lane infrastructure, 37
　geofencing, 38
　infrastructure, 37
　locking mechanisms, 39
　low-cost parking, 37
　major shared operators, 38
　mandatory parking spots, 38
　mobility on demand, 85–89
　non-car mode of transport, 36
　place to park, 38–41
　place to ride, 41–45
　political dynamic, 28
　protected lanes, 25–27
　road geometry, 33, 34

scooters
  economics, 49–50
  equity zones, 63–64
  fleet size and parity, 53–55
  labor costs, 62–63
  number of operators, 50–53
  operating areas, 55–57
  parking compliance and fines, 64–66
  parking requirements, 37
  permit fees, 60–62
  permit length, 57–58
  regulations, 59
  rider restrictions, 66–68
  selection criteria, 58–59
  sidewalk riding, 41, 68–69
  technology and urbanism, 33
  transit bus, 33, 73
    boarding/payment, 81–83
    bus lanes, 77–80
    competition, 74–75
    frequency, 74, 76–77
    operating cost, 76
    pricing, 75
    reliability, 74, 76–77
    routes and times, 74
    schedule planning, 75–76
    signal priority, 80–81
    speed, 74, 77
  transportation budgets, 37
  transportation planners, 37
  travel modes, 34
  unnecessary fees and burdensome requirements, 37
Mobility on demand
  Access Services ride, 88
  coverage routes, 86–87
  customer service, 89
  fixed routes, 87
  ridership-focused routes, 86
  service credits, 88
  shuttle buses, 85
  trade-off, 86
  transit agencies, 86, 87
Multimodal transportation, 29

N
National Association of City Transportation Officials (NACTO), 12–14, 17, 20, 39, 51–52, 54
Neighborhood rules, 38

Non-car modes, 8
North American Bike Share Association (NABSA), 35

O
Operational cost, 55, 58, 68

P
Parking compliance technology, 58–59, 64–66
Parking requirements, setting
  existing infrastructure, 38
  pedestrian traffic, 38
  population and local activity, 38
Policy change, 27–29
Policy support, 20
Pricing, 7, 8, 12, 13, 17, 21, 60, 75

R
RFP process, 51, 52
Ride-hailing, 23
Rider education, 42–43
Ride services, 6
  availability of, 24
  regulation, 7
Road pricing, 11
  economic solutions, 15
  income-based discounts, 21
  induced demand, 14
  mechanisms, 7
  traffic in cities, 13–14
  vehicle miles traveled, 12–13

S
San Francisco Municipal Transportation Agency (SFMTA), 13, 16
Scooters
  economics, 49–50
  equity zones, 63–64
  fleet size and parity, 53–55
  labor costs, 62–63
  number of operators
    NACTO, 51–52
    open markets, 50–51
    partnership with operator, 52
    permit limits, 51
    recommendations, 51
    single-operator contracts, 52
  Superpedestrian, 52–53

  operating areas, 55–57
  parking compliance and fines, 64–66
  permit fees, 60–62
  permit length, 57–58
  regulations, 59
  rider restrictions, 66–68
  selection criteria, 58–59
  sidewalk riding, 41, 68–69
SFMTA report, 29, 35, 42, 78, 81
Shared micromobility operators, 33, 35, 43, 60, 66, 67, 69
Shared scooter programs
  in Paris, 45–47
  political backlash, 47
  regulatory framework, 47
Social Bicycle's Smart Bikes, 49
Speed limits, 67
Streetfight, 28
Street parking, 29
Superpedestrian, 52–53

T
Tactical transit lanes
  BRT, 78–79
  car drivers, 79–80
  definition, 77
  investment, 77–78
  self-enforcement, 79
  traffic crashes, 79
Taxis, 6, 15, 18, 34
TechCrunch, 45, 46
Traditional infrastructure undertakings, 43
Transit agencies, 86, 87, 89
Transit bus, 73
  boarding/payment, 81–83
  bus lanes
    BRT, 78–79
    car drivers, 79–80
    definition, 77
    investment, 77–78
    self-enforcement, 79
    traffic crashes, 79
  competition, 74–75
  frequency, 74, 76–77
  operating cost, 76
  pricing, 75
  reliability, 74, 76–77
  routes and times, 74
  schedule planning, 75–76
  signal priority, 80–81
  speed, 74, 77

Transit Center, 95
"Transit First" policies, 33, 34
TransitScreen, 83
Transit Signal Priority (TSP), 80–81
Transportation
  agency, 35, 36
  car, 3
  costs for, 90
  fare policies, 94
  private operators, 89
  public budgets, 90

Transportation-related complaints, 6
Transport for London (TfL), 44
Trips per vehicle per day (TVD), 51, 53, 62

U
Uber, 7, 61, 75, 87, 89, 92
Uber blog, 14–15
UCLA Institute of Transportation Studies, 78

Ultra Low Emission Zone (Ulez), 18, 56, 57
Unmetered street-parking spaces, 39
US Department of Transportation, 13, 16

V
Vehicle docking station, 61
Vehicle miles traveled (VMT) fee, 12–13, 21
Via, 88, 92

About the Author

Michele Kyrouz is a lawyer, writer, and podcast host based in the San Francisco Bay Area. She has been involved in the new mobility community since 2017 as host of the *Smarter Cars* podcast, author of numerous articles on mobility policy, and frequent contributor at conferences including Micromobility Europe and Micromobility World. Michele has a B.A.  in Political Science from U.C. Berkeley and a J.D. from Columbia University School of Law. She is a former partner at Latham & Watkins with 20 years of experience as a regulatory lawyer. She most recently served as general counsel for the micromobility company Superpedestrian.